Volume 14

Make:

technology on your time™

SPECIAL SECTION OPTICS

ON THE COVER: Dave Sims enjoys the buggy's-eye view, while daughter Ellie climbs to safety. Photographed by Robyn Twomey and styled by Alex Murphy and Sam Murphy.

Columns

DOUBLE VISION: Eric Kurland shoots a 3D video with his home-built dual camera rig.

+ makezine.com
Visit for story updates and extras, Weekend Project videos, podcasts, forums, the Maker Store, and the award-winning MAKE blog!

Vol. 14, May 2008. MAKE (ISSN 1556-2336) is published quarterly by O'Reilly Media, Inc. in the months of March, May, August, and November. O'Reilly Media is located at 1005 Gravenstein Hwy. North, Sebastopol, CA 95472, (707) 827-7000. SUBSCRIPTIONS: Send all subscription requests to MAKE, P.O. Box 17046, North Hollywood, CA 91615-9588 or subscribe online at makezine.com/offer or via phone at (866) 289-8847 (U.S. and Canada); all other countries call (818) 487-2037. Subscriptions are available for $34.95 for 1 year (4 quarterly issues) in the United States; in Canada: $39.95 USD; all other countries: $49.95 USD. Periodicals Postage Paid at Sebastopol, CA, and at additional mailing offices. POSTMASTER: Send address changes to MAKE, P.O. Box 17046, North Hollywood, CA 91615-9588. Canada Post Publications Mail Agreement Number 41129568. CANADA POSTMASTER: Send address changes to: O'Reilly Media, PO Box 456, Niagara Falls, ON L2E 6V2

Make: Projects

Living Room Baja Buggies
With wireless cameras, these radio-controlled racers give you virtual reality telepresence. By John Mouton

96

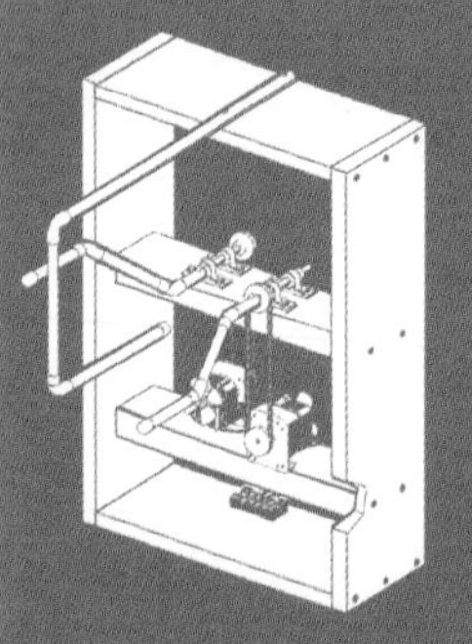

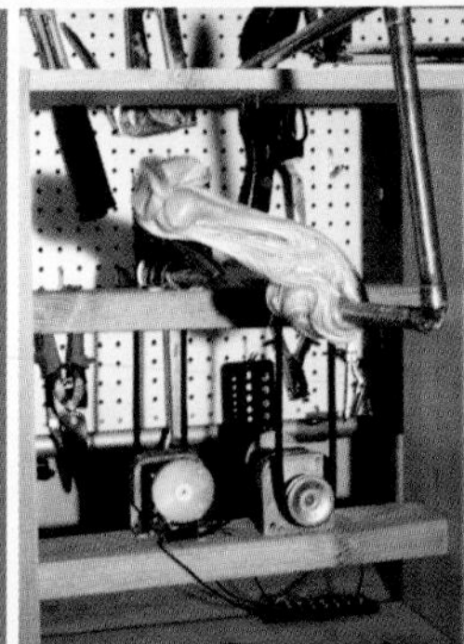

Taffy Machine
Make a simple mechanism that pulls delicious candy while it stretches the limits of multidimensional math. By William Gurstelle

106

Pixelmusic 3000
Re-create Atari's classic music visualizer by plugging this box into any TV and audio source. By Tarikh Korula

114

Solar Power System
How to use solar panels to supplement your home or workshop electricity needs. By Parker Jardine

160

oject: Solar Panel Backpack

eated by: Todd Vanderlin

RadioShack.
Do Stuff℠

rts:

 4x +  +  +   +  +  + 

otes:

Last time I went hiking, both my wireless phone and iPod were low on power. There was no way to keep them charged. So here's my idea: Solar panels soldered to a USB adapter. Then I just attach them to my backpack, and BAM! I've got myself a solar panel backpack. Just stop by your neighborhood RadioShack to start constructing your own invention.

Volume 14

Make:

technology on your time™

READ ME: Always check the URL associated with a project before you get started. There may be important updates or corrections.

Maker

28

COCOA PUNKS: **TCHO's Timothy Childs (right, with Matt Heckert) and his rebuilt chocolate-making system.**

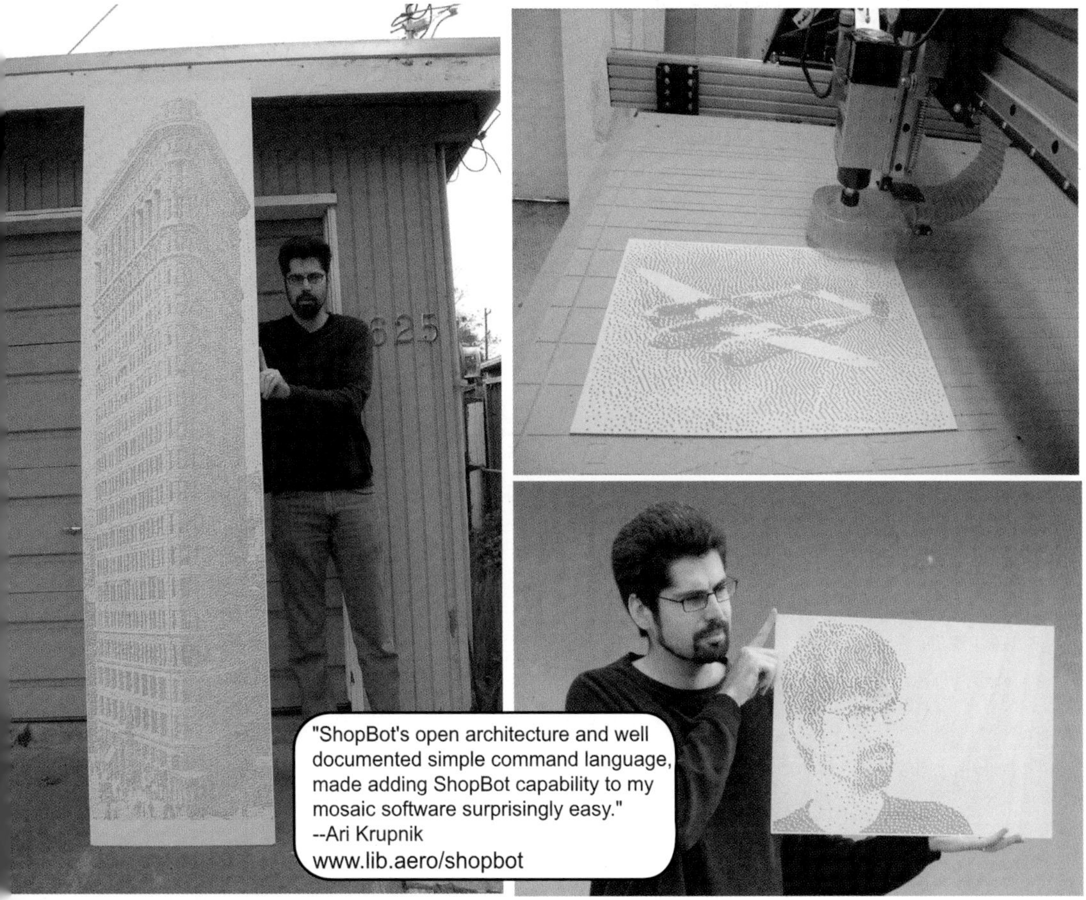

625
"ShopBot's open architecture and well documented simple command language, made adding ShopBot capability to my mosaic software surprisingly easy."
--Ari Krupnik
www.lib.aero/shopbot

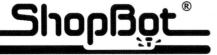

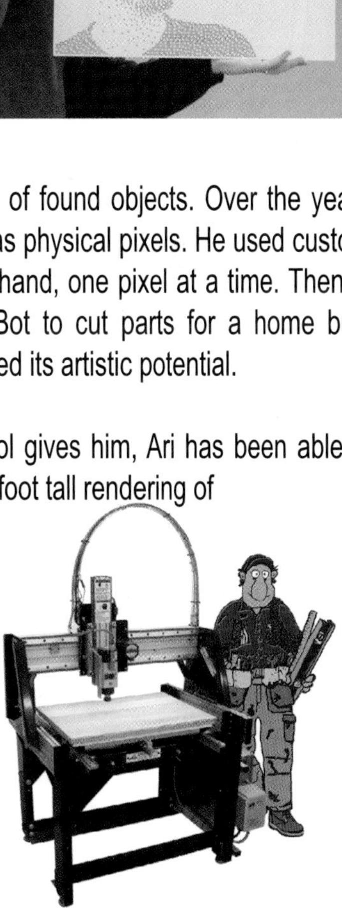

PUBLISHED BY O'REILLY MEDIA, INC.
Tim O'Reilly, CEO
Laura Baldwin, COO

Visit us online at makezine.com
Comments may be sent to editor@makezine.com

For advertising inquiries, contact:
Katie Dougherty, 707-827-7272, katie@oreilly.com

For event inquiries, contact:
Sherry Huss, 707-827-7074, sherry@oreilly.com

Customer Service cs@readerservices.makezine.com
Manage your account online, including change of address at:
makezine.com/account
866-289-8847 toll-free in U.S. and Canada
818-487-2037, 5 a.m.–5 p.m., PST

EVEN GREENER: NEW SOY INK!
MAKE is printed on acid-free, recycled paper containing 30% post-consumer waste, with soy-based inks containing 22%–26% renewable raw materials. Subscriber copies of MAKE, Volume 14, were shipped in recyclable plastic bags.

PRINTED WITH SOY INK

PLEASE NOTE: Technology, the laws, and limitations imposed by manufacturers and content owners are constantly changing. Thus, some of the projects described may not work, may be inconsistent with current laws or user agreements, or may damage or adversely affect some equipment.

Your safety is your own responsibility, including proper use of equipment and safety gear, and determining whether you have adequate skill and experience. Power tools, electricity, and other resources used for these projects are dangerous, unless used properly and with adequate precautions, including safety gear. Some illustrative photos do not depict safety precautions or equipment, in order to show the project steps more clearly. These projects are not intended for use by children.

Use of the instructions and suggestions in MAKE is at your own risk. O'Reilly Media, Inc., disclaims all responsibility for any resulting damage, injury, or expense. It is your responsibility to make sure that your activities comply with applicable laws, including copyright.

Printed in the USA by Schumann Printers, Inc.

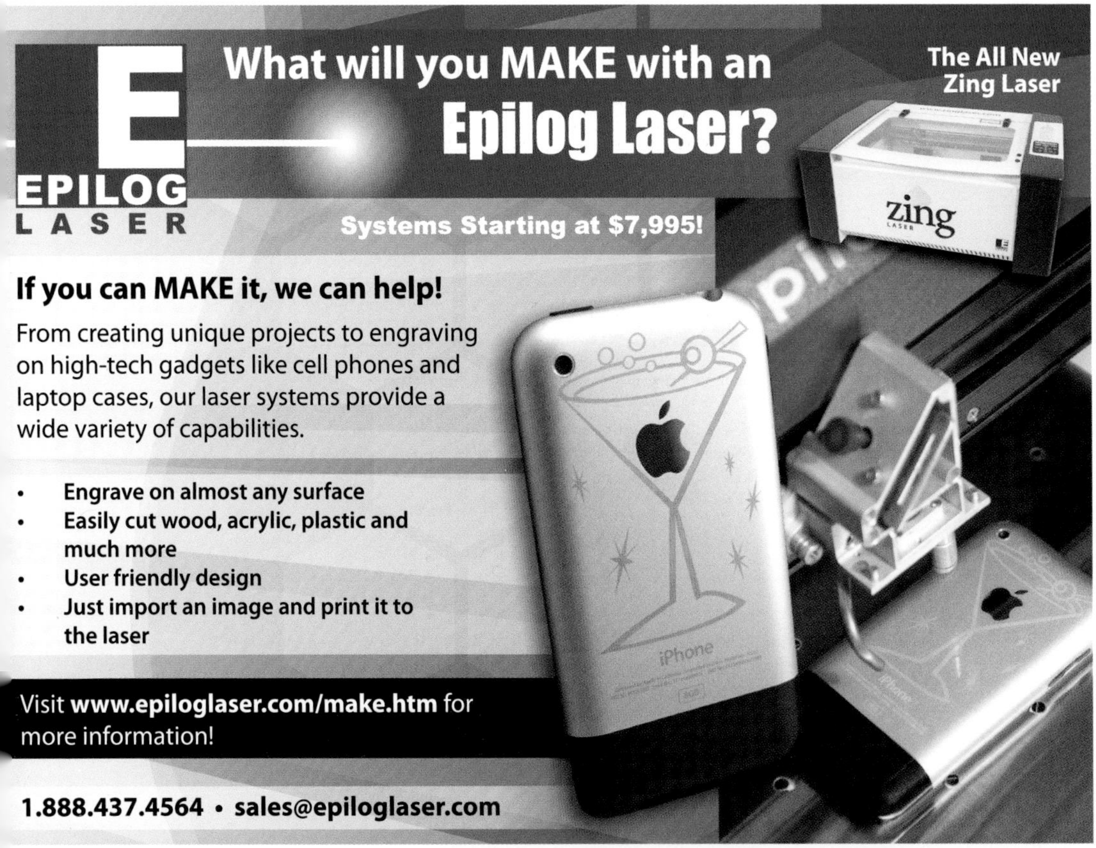
What will you MAKE with an
Epilog Laser?
EPILOG
LASER
The All New
Zing Laser
zing
Systems Starting at $7,995!
If you can MAKE it, we can help!
From creating unique projects to engraving on high-tech gadgets like cell phones and laptop cases, our laser systems provide a wide variety of capabilities.
• Engrave on almost any surface
• Easily cut wood, acrylic, plastic and much more
• User friendly design
• Just import an image and print it to the laser
Visit www.epiloglaser.com/make.htm for more information!
1.888.437.4564 • sales@epiloglaser.com

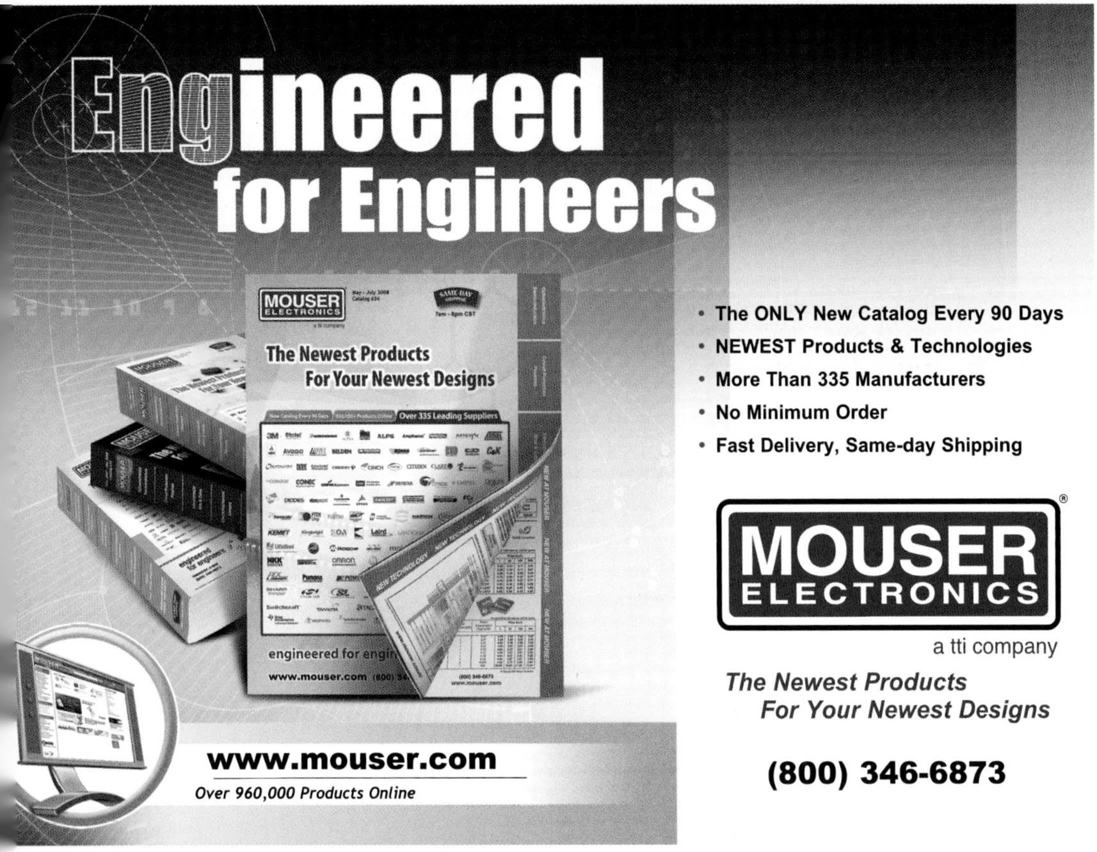
Engineered
for Engineers
The Newest Products
For Your Newest Designs
• The ONLY New Catalog Every 90 Days
• NEWEST Products & Technologies
• More Than 335 Manufacturers
• No Minimum Order
• Fast Delivery, Same-day Shipping
MOUSER
ELECTRONICS
a tti company
The Newest Products
For Your Newest Designs
(800) 346-6873
www.mouser.com
Over 960,000 Products Online

Contributors

"A fan of the noble undertaking, the success snatched from the jaws of defeat, and the spectacular failure," **Gever Tully** (*Design Build Prize*) is "chaotically productive, fanatically messy, and prone to flights of fancy." While he once wrote code for 72 hours straight, the founder of Tinkering School now spends considerably more time working with his hands. He lives in Montara, Calif., with his "lovely wife and collaborator, Julie Spiegler," and their dog, Daisy Love Cracker. Proving that *committed tinkerer* is not an oxymoron, he admits that "my favorite tool is my brain, my second favorite are my hands, and after that it's a toss-up between the plasma torch, the laser cutter, and the pocketknife."

As a kid, **Tarikh Korula** (*Pixelmusic 3000*) would hack (and break) his video games, alarm clocks, joysticks, and CD players. As an adult, he finds himself trying to explain why he still does those things. He's managed to further complicate matters by getting his hands dirty in sound art, media reform, and small business. In 2004, Tarikh and Josh Rooke-Ley founded Uncommon Projects with the intent to make a living by answering some of these conundrums. uncommonprojects.com

Amy Crilly (*3D Movies* photography) is a Jersey girl by way of San Diego. She's been studying photography ever since she discovered her love for it. Amy earned a B.A. from California College of the Arts in the San Francisco Bay Area and began her professional career as a photo editor at *Wired* magazine. After three years producing shoots for other photographers, she decided to give it a go herself, so she packed up her husband and moved to Los Angeles, where she now shoots for numerous publications and clients. amycrilly.com

When **Brad Graham** and **Kathy McGowan** (*Evasive Beeping Thing*) aren't inventing, consulting, or knee-deep in electronics, robotics, and bike-building, they enjoy the outdoors, especially "biking, camping, fishing, and concocting Atomic Zombie and Evil Genius ideas by the campfire." They're currently working on a heavy-duty outdoor security robot, a retro arcade game for hobbyists, and more Atomic Zombie bikes: recumbents, trikes, choppers, and an electric motorcycle. Robotics advice? "Never build a bot you can't outrun!" atomiczombie.com

John Mouton (*Baja Buggies*) is an applications engineer with Microchip's Security, Microcontroller, and Technology Development Division (SMTD). He ensures that Microchip's products meet silicon validation specifications, works with customers on design projects, and helps to develop and demonstrate product applications. He spent four years in the U.S. Air Force and five years in college, completed a co-op program with Advanced Micro Devices, and worked with high-voltage power products at 3M. His advice to engineers? "Be versatile, have fun with everything you do, and always continue learning. Because when you're green, you grow, and when you're ripe, you're rotten."

Ed Troxell (editorial intern) is a communications major at Sonoma State University who signed up for a photojournalism class "not knowing what to expect," only to become photo editor and a contributing photographer for the school newspaper. He's interned at the *North Bay Bohemian*, where his very first assignment was a hit, and had his photographs published in the *San Francisco Chronicle*. When he's not fact-checking or proofing articles at MAKE, Ed likes scary movies, traveling, heading out to clubs, and BBQing. He just built a backyard bar from scratch.

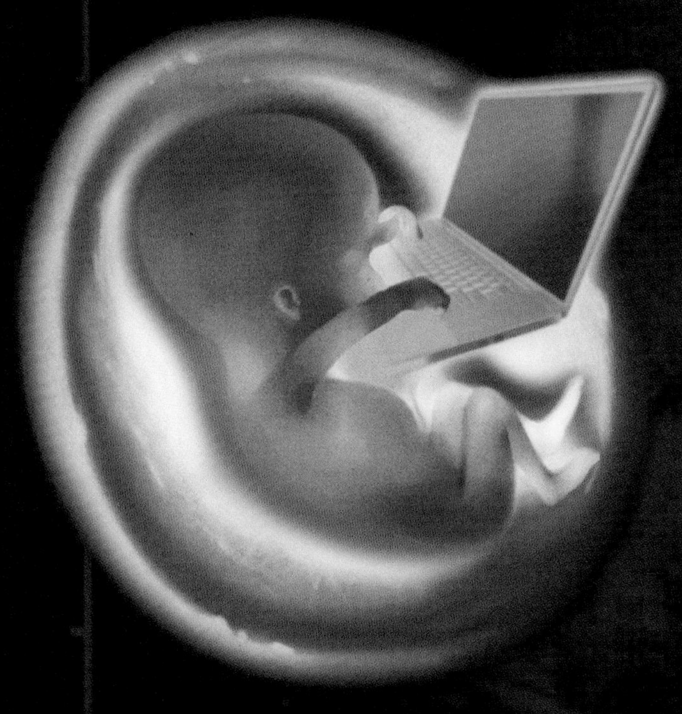

BY MARK FRAUENFELDER

Make Like Picasso

Recently I came across a quotation about making things, attributed to Pablo Picasso. I found it in a wonderful, hand-lettered how-to book from 1973 called *Nomadic Furniture 1: How to Build and Where to Buy Lightweight Furniture That Folds, Inflates, Knocks Down, Stacks, or Is Disposable and Can be Recycled*, by James Hennessey and Victor Papanek.

Here's what Picasso said: "When you make a thing, a thing that is new, it is so complicated making it that it is bound to be ugly. But those that make it after you, they don't have to worry about making it. And they can make it pretty, and so everybody can like it when the others make it after you."

Picasso was right, and the authors of *Nomadic Furniture* were wise to quote him, because a lot of the furniture pieces in the book are functionally clever, but they're eyesores. That's not surprising for first generation, proof-of-concept prototypes. The authors included the quote as a challenge to the reader: we worked hard to design this easy-to-build furniture, now it's up to you to improve on what we've create and make it look pretty.

Some of the projects in MAKE aren't beautiful to look at. They do what they're supposed to do, but they lack aesthetic appeal. I know that many makers are more interested in function than in form, because the challenge of getting something to work the way you imagined it can be an all-consuming activity. That's great, but when you create something that's amazing, it makes sense to honor it by endowing it with physical appeal, too.

The object shown here is an example of an attractive package for a neat homebrew gadget called the Multari (retroactive.be/multari). It's a handheld Atari 2600 clone built and designed by a teenager named Marshall H. from Kansas. Marshall packaged the circuitry for the Multari in a vacuum-formed styrene plastic case, and it looks terrific. (*To create your own vacuum-formed 3D parts, check out MAKE, Volume 11, page 106, for instructions on setting up a Kitchen Floor Vacuum Former.*)

Throughout his life, Picasso never stopped challenging himself to learn new ways of doing things.

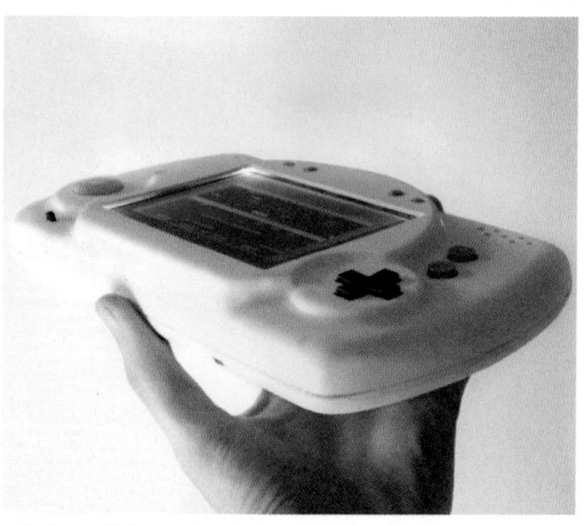

Photograph by Marshall Hecht

When you create something that's amazing, it makes sense to honor it by endowing it with physical appeal, too.

Why not challenge yourself by learning how to make your creations look better?

Here are three more pieces of good advice often attributed to Picasso (found on paintalicious.org) about becoming a better maker:

1. "He can who thinks he can, and he can't who thinks he can't. This is an inexorable, indisputable law."

2. "I am always doing that which I cannot do, in order that I may learn how to do it."

3. "Action is the foundational key to all success."

Have you come up with a way to make your projects look pretty? We'd love to see them. Please show us at makezine.com/14/welcome.

Mark Frauenfelder is editor-in-chief of MAKE.

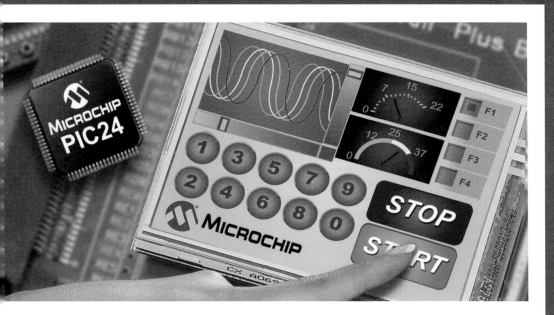

BY CORY DOCTOROW

Too Much Time on My Hands

I became a dad in early February, and during the run-up to The Day, I worked my butt off, getting as much work in the can as I could before my life got taken over by the fruit of my loins.

All that hard work paid off. One day, I looked around and realized that it was all done. Everything. I'd been ruthless about telling people that I couldn't even consider their new projects until I had a handle on fatherhood, and I'd caught up on all the assignments and to-dos that had filled my long, long list. For the first time since I dropped out of college, I found myself with nothing — *nothing*! — to do.

Whoa. What a feeling.

It was a little scary at first, but after about half an hour's thought in my new office (my old office is now the baby's room, so I had to rent a place to locate all my junk and carve out some space where I could write that wasn't the living room sofa), I realized that there was one thing I really wanted to do.

I wanted to paint a D&D miniature.

I discovered Dungeons & Dragons and similar tabletop games when I was about 10, and I was immediately taken with painting minis. I was never very good at it, but I derived immense, all-consuming pleasure from the activity, which became a kind of meditation. So I walked down to Covent Garden here in London and dropped in on the Orc's Nest, the RPG store whose windows are always filled with beautifully painted minis. There, I bought brushes and paints and a selection of tiny lead monsters.

Then I sat down to paint. When I finished my first mini — two days later — I saw that it had all come back to me, that bone-deep satisfaction I'd derived all those decades ago. I put a pic of my little winged vampire critter up on Flickr and got a lot of good feedback — and a couple of remarks that contained the phrase "too much time on your hands."

Now that's a thoughtless and immensely cruel way of negating the pleasure that the subject has derived from following his passions. It takes in the sweep of someone's gnarly physical meditation and grinds it into paste.

All creative endeavor begins with just fooling around, not doing much of anything, just noodling and letting the different parts of your mind talk to each other. Science and art and invention spring forth when we do the unexpected, and so coax our brains into letting imaginative combinations of ideas jangle together. Working with your hands, taking a walk, singing a song, doing a drama exercise, building something, designing something, painting — they engage parts of our brains in ways that we're probably not used to.

All creative endeavor begins with just fooling around, not doing much of anything.

So I don't care if you're scouring yard sales for Beanie Babies, overclocking your PC, speedrunning *Super Mario*, landscaping your garden, or building a trebuchet out of a fallen telephone pole. I don't care if the end product works or not. I don't care if it's too ugly to look at.

Here's what I care about: did you follow your weird? Did you get into that blissed-out concentration state that great athletes and musicians and artists find themselves in? Did you go to a place where your mind was able to talk to itself without the endless chatter of the million billion grocery items and nagging doubts?

If you got there, you're winning the game of life.

Photograph by Alice Taylor

Cory Doctorow lives in London, writes science fiction novels, co-edits Boing Boing, and fights for digital freedom.

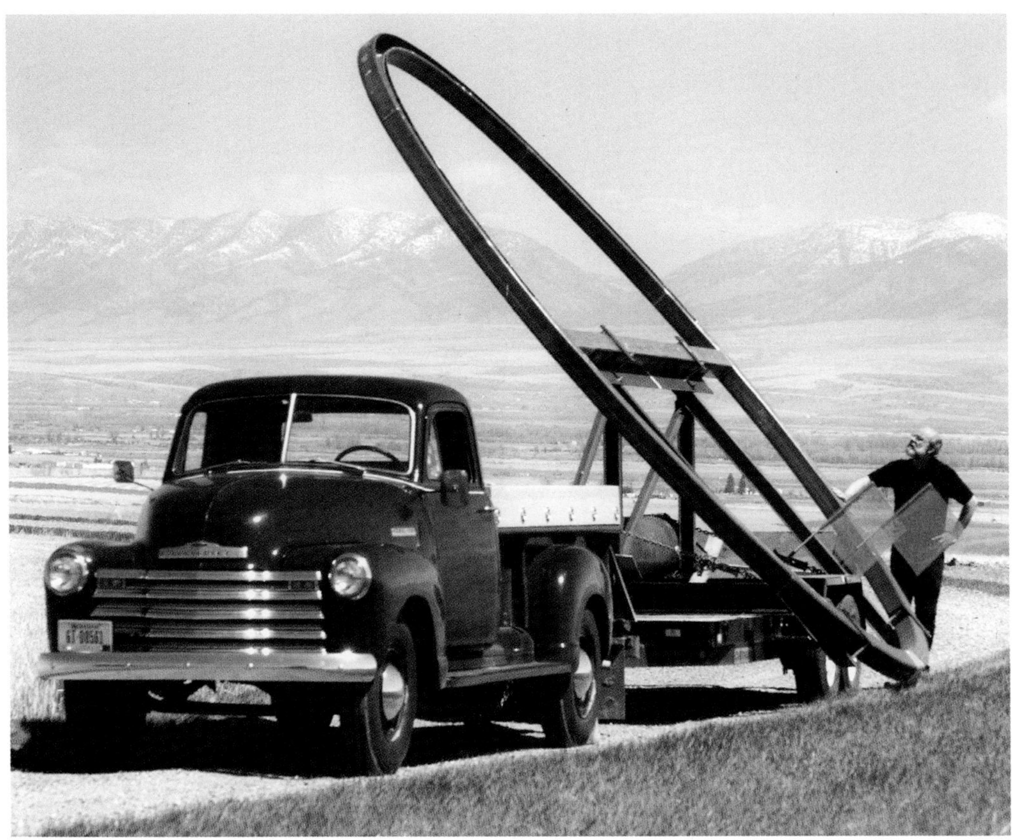

Earth, Wind, Inspire

Gary Bates grew up plowing the fields of his grandfather's farm in tiny Manhattan, Mont., where he now lives. Making passes on the tractor each day bored the young Bates, but he did enjoy lying on the grass and watching the windmills. And it was there that he found the inspiration for his kinetic sculptures.

In his early 20s, Bates began building large, wind-powered structures from recycled farm machinery. He placed these sculptures on the edge of the field so he could watch them while he drove the tractor, sometimes looking at them from a mile away.

Today, a telescope points from Bates' living room to his 1986 sculpture *Lunar Ketcherschmitt*, a 14-foot-high piece marking the edge of his property. *Ketcherschmitt* is made of an old steel boiler cut in two, with one 2,300-pound half spinning atop the other. Engineers from Stanford University have visited to study how the wind can start the heavy top half spinning, but they remain baffled. Bates doesn't necessarily understand it either. "I don't know why it works," he admits, "but I'm happy that it works."

Like many of Bates' sculptures, *Ketcherschmitt* makes visible the pulse of the environment. Each one reacts to some natural force — in this case, wind — and transmits the information in a visual way. Bates peers through the lens to *Ketcherschmitt* each morning to see what the weather might be like. Similarly, an engineering professor at Montana State University watches the spin of Bates' *Wind Arc* from his office window to determine whether it's too blustery to ride his bike home.

Bates' next public work, *Rain Scale*, will be installed this year at Green River Community College in Auburn, Wash. Bates will perch an 18-foot-wide horizontal ring of stainless steel atop a 25-foot-high arch. Three-eighths of an inch of rain, or 29 pounds of water, will set the 2,000-pound ring into see-sawing motion for almost an hour, depositing water into the pond below. It's sure to be a glorious sight — just remember to bring your umbrella.

—Linda Permann

>> **Monumental Kinetic Sculpture:** sculptorgarybates.com

Photography by Jens J. Selvig III

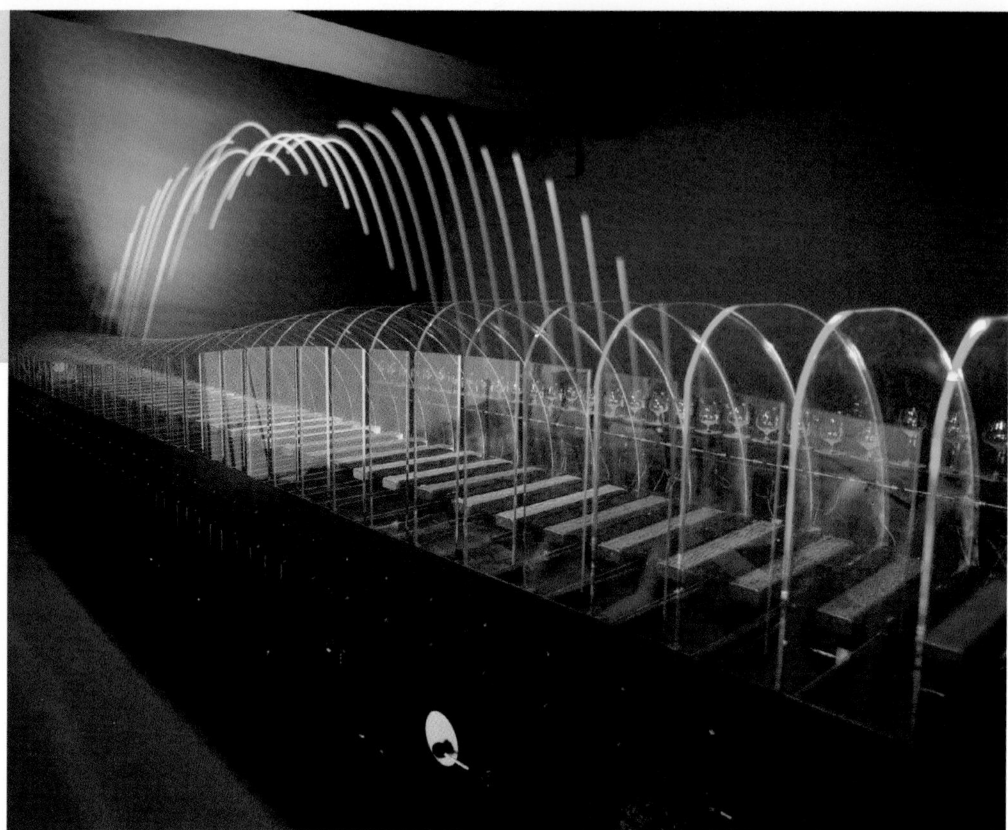

Musical Engineerity

Want robots to be musical, creative, and expressive? Better brush up on your engineering. Musician/roboticists **Dan Paluska** and **Jeff Lieberman** constructed a web-connected "robotic mechanical orchestra" that plays a marimba by firing rubber balls out of a cannon, strikes traditional percussion instruments, and also rubs mechanical fingers along wine glasses. The machine, *Absolut Quartet*, uses artificial intelligence to creatively riff on melodies composed remotely by users on the web.

"At the core, the machine is just motors, metal, and software," say the MIT alums. "However, the design of these elements gives the whole machine a 'personality' and this is what allows a creative dialog to exist between the machine and the online user."

Of course, that dialog can't just work once — it has to work over and over again. The guys wanted the technology to "disappear," leaving a purely creative experience. But that meant making 3,000 custom parts and 10,000 stock parts work in harmony.

And then there are the 500,000 custom rubber balls firing a 4-meter arc onto the keys.

"For any reasonable maintenance, this can only fail roughly 1 in 10,000 times," the duo explains. They tried four fundamentally different shooting mechanisms before they found one that worked — springs and a rotating arm.

They then consulted an engineer to settle on magical, maintenance-solving ingredients such as polyethylene glycol dimethacrylate, which they used to make the suede fingers resonant. But they also needed the skills of a professional glass harpist so they could get 35 tuned wine glasses.

"Being both musicians and roboticists, we have always been interested in combinations of the two," say Paluska and Lieberman. In the finished work, centuries-old percussion and glass armonicas meet modern industrial robotics. Musician/inventor Benjamin Franklin, who built the first glass armonica, would have been proud. *—Peter Kirn*

Absolut Quartet: absolut.com/absolutmachines

>> The Build: bea.st/sight/absolut

Photograph by Jeff Lieberman

Pedal Pure

Providing clean water for all could be as easy as riding a bike. Or a trike, if Aquaduct has an influence. Winner of the 2007 Innovate or Die pedal power competition, the Aquaduct Mobile Filtration Vehicle stores, transports, and purifies water as it goes.

"We came up with ideas ranging from ways to clean up oil spills in the Bay to how to boil an egg," says **Brian Mason**, one of Aquaduct's five designers, all of whom work at the Palo Alto, Calif., design firm IDEO. "But we kept coming back to the need for clean water in the developing world."

More than 1 billion people lack access to clean water. Trekking miles to fetch it can take hours, and boiling it for sanitation uses precious resources. Aquaduct reduces the strain of hauling water, and its closed system prevents contamination.

Simply ride to a source, fill the 20-gallon storage tank — a day's supply for a family of four — and pedal home, filtering all the way. Clean water drains into a removable container that can be brought indoors. Once that's empty, the pedals can be disengaged from the wheels and the vehicle ridden in a stationary position to filter the rest.

"The answers are out there," says another of Aquaduct's designers, **Paul Silberschatz**. "Through design and innovation, we can find simple solutions to even the most challenging problems."

The team, including **Adam Mack**, **Eleanor Morgan**, and **John Lai**, used 2D and 3D modeling to help them modify a Miami Sun tricycle frame, custom-build a peristaltic pump that draws water through a simple filter, and cover surfboard foam in fiberglass to round out the body. Simple sanding and automotive paint finished the job, explains Silberschatz, who, luckily, used to build race cars.

The IDEO crew donated the contest's $5,000 purse — along with a $10,000 match from sponsors Google and Specialized — to Kickstart, a nonprofit that develops and markets new technologies in Africa. But they did ride away with something: each member got a brand-new urban commuter bicycle called the Globe. —*Megan Mansell Williams*

Aquaduct in Action: makezine.com/go/aquaduct

Photograph by Nicolas Zurcher

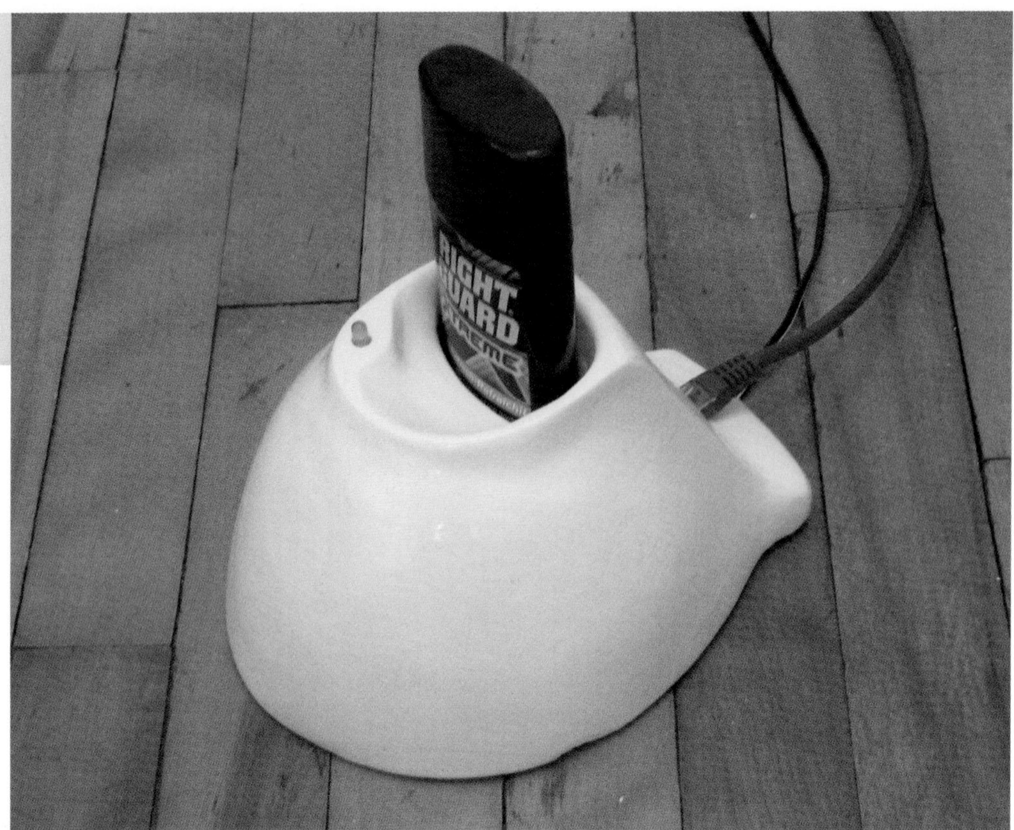

Photograph by Laurier Rochon

Status: Stinky

Bridging the gap between real-world hygiene and the social networks taking over the virtual world, iamclean.org offers up an intriguing way to keep tabs on someone's level of body odor.

Laurier Rochon and **Marc Beaulieu** built a web-enabled deodorant docking station, originally called *Zero Privacy*, as a whimsical critique of social networking sites and their related privacy issues.

The dock senses when Rochon's deodorant stick is removed, keeps track of how long it's out, and then instantly transmits that information to the web for the whole world to see. Anyone online can tell on which days and for how long Rochon has applied his deodorant. If this seems like a creepy invasion of privacy and makes you feel like screaming "too much information!" you're getting their point.

The deodorant dock was designed as a class project in Vincent Leclerc's course Physical Computing and Tangible Media, at Concordia University in Montreal. Rochon and Beaulieu wanted to use the same social systems they say are causing people to give up too much privacy, believing that parody is better at raising awareness than ranting.

From idea to finished product, it took the duo four weeks to complete the dock. The website lays out their plans, materials, and schematics, so others can build on their work without struggling through the same challenges, like trying to figure out the XPort Ethernet server's communication protocol so it talks to the internet the way they wanted it to.

They placed a high importance on making the dock look almost overdesigned, like a commercial product, to add to its believability and humor. It's got a status LED that goes on when the deodorant is removed, and an Ethernet jack for easy connection to the web. It runs on AC power or AA batteries, and the future addition of a wireless connection will make it even more portable.

By combining creative making, software hacking, and satire of the idea of online "status," the project is helping to keep Rochon clean, one day at a time.

—Bruce Stewart

» **Zero Privacy:** iamclean.org

Photograph courtesy of Liz Cohen

Auto Erotic

Five ratchet-wielding years, one East German automobile, and several coats of bikini wax — for **Liz Cohen**, it's been a long, sticky trip. The performance and documentary artist has built one of the most improbable custom cars in the country, and has the pictures to prove it.

Most recently on display at Arizona's Scottsdale Museum of Contemporary Art, Cohen first threw her *Bodywork* project into gear by cutting a 1987 Trabant 601 Deluxe in half. Then she ditched its 26-horsepower, 2-stroke motor for a 305 small-block engine nine times more powerful and rebuilt the vehicle from the wheels up.

The result is a former-Communist lawnmower that becomes a 1973 Chevy El Camino muscle car at the flip of a few switches, "just like some kind of incredible Frankenstein," Cohen says.

Thanks to hydraulics, the *Trabantamino* grows 14 inches in height and 6 feet in length when it goes Camino. Coiled Teflon brake and fuel lines and dovetailing fiberglass side panels extend and contract as it changes shape. The specially built drive shaft telescopes four times, and still runs at full speed without vibrating. "It's this weird thing that doesn't fit in but figures out a way to get accepted," she says.

Which is an apt description of the artist's whole journey. When she began working at Scottsdale's Elwood Body Works, Cohen had little experience with tools and was entering the mostly male world of custom car builders and owners. But rather than try to pass as a dude, she buffed up her own body and became a bikini model. Her calendar of sexy shots will be out this summer.

What does the custom car scene think of Cohen and her wild ride? "Auto shop owners get so excited when I tell them about the project that they give me discounts and things for free," she says. But the real test comes this summer, when she'll pull the *Trabantamino* into lowrider competitions across the country, jump in front in a skimpy outfit, and see if her monster wins any prizes.

—Eric Smillie

» **Liz Cohen:** myspace.com/trabantamino

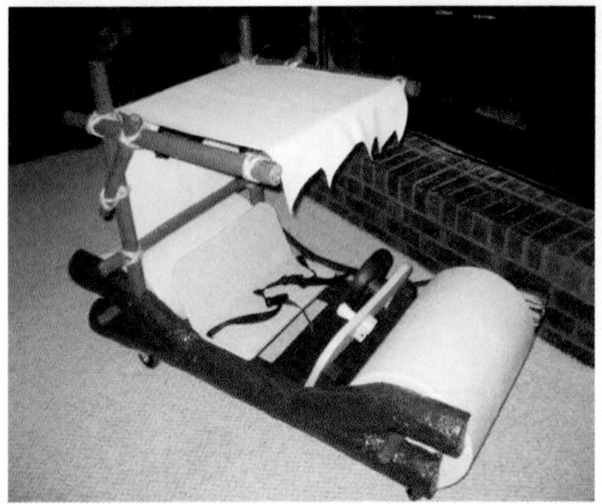

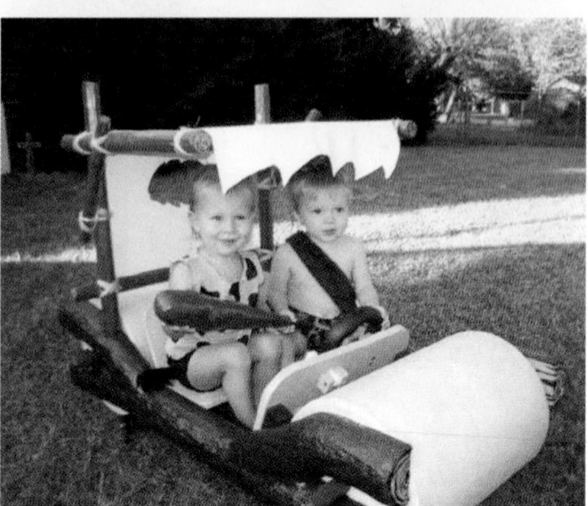

Yabba Dabba DIY

Bill LeMaster, a 44-year-old retired Air Force technician living in Montgomery, Ala., enjoys a great many hobbies, ranging from arts to electronics. He also enjoys his family, his most beloved passion. Although, if you ask his wife, she'd probably say collecting hobbies is his biggest passion.

Last Halloween, when he heard his grandkids were dressing up as Pebbles and Bamm-Bamm from the Flintstones, LeMaster volunteered to build them the Flintstone car to perfectly complete the look (and to make sure he got to join in the fun). "I just wanted my grandkids to have the most awesome costume in Montgomery," he recalls.

He started the build six weeks before Halloween, allowing himself plenty of time to get the project rock solid. Once the car was done, the kids were all set to shuffle their feet down the street — Yabba-Dabba-Doo! His grandkids were happy, the neighbors were amazed, and LeMaster was satisfied with the outcome of the project.

People continue to ask him where he bought the car and if he'll bring it out of the house. In response, LeMaster has posted a how-to, comprised of step-by-step instructions on making your own Flintstone car, on the Instructables website.

"I originally documented the construction to simply capture the memories of the build, but it just so happened that I came across Instructables shortly after I started so I decided to enter the contest. I figured people might be interested in some of the techniques I came up with," he explains.

LeMaster says it took him about three weeks to build the car, with most of the work done on Saturdays. The bulk of the material he used was scrap wood that he found lying around the house. Swim noodles and bondo glass both helped in the construction.

All in all, the project cost him about $100, and LeMaster says it was worth every penny. "This was nothing compared to the priceless expressions on my grandkids' faces when they saw the car."

—Ed Troxell

>> **Flintstones Car How-To:** makezine.com/go/flintmobile

Photography courtesy of Bill LeMaster

Dirty Car Art

Scott Wade of San Marcos, Texas, thought he could do better than write "Wash Me" on the backside of a dusty car. He started drawing caricatures. His father was a cartoonist of sorts and had taught him to draw funny faces. It was Wade's idea to make a dirty car window his canvas.

"For the last 20 years living on a dirt road," he says, "there's always dirt on my car."

With the sun baking it, the dirt takes about two weeks to form a stable work surface. Wade began, like anyone else, by using his finger, and then tried popsicle sticks. To introduce shading, he decided to use brushes. Over time he developed a range of techniques, which included using plants and rubber paint-shaper tools.

Wade particularly likes the dirt of central Texas, where crushed limestone mixed with clay serves as a road base.

"It makes the perfect dirt," he says. "It's very light-colored and the contrast is great against the dark shadow inside the car."

As he got more requests to create his Dirty Car Art in public, he realized that he had to figure out how to dust up a car himself. Now, he can prepare a car in minutes using a light coating of oil and pyrolite, a less toxic alternative to fuller's earth.

At the Austin Maker Faire in 2007, Wade dusted up his Toyota and created *Monsters from the Movies*, featuring the Phantom of the Opera, Dracula, Frankenstein, and the Wolfman. The next day he painted a tribute to Willie Nelson that included Waylon Jennings. "After a good rain," he says, "it appears to wash off, but in a couple days it comes back in a ghostly form."

Recently, he was asked to draw Biff Henderson for the David Letterman show. In addition to portraits, he enjoys dusting up the old masters. "I have this grandiose idea of parking cars all the way up the ramp of the Guggenheim Museum and painting in dirt reproductions of the pieces that are on the wall next to it."

—Dale Dougherty

» **Dirty Car Art:** dirtycarart.com

Photograph by Kim Dow

The Power of Things

Recently I have been preoccupied with pretty much only one thing, and I think it is genuinely the challenge of our times: energy and the environment. In engaging with this issue, you have to think about how we harness energy (energy isn't created, but is instead converted to different forms), how we use power, and how we affect the environment. I've done a lot of calculations of humanity's energy use, and how we might make renewable energy technologies that can meet that challenge, but it's a fairly impersonal analysis.

To make it more immediate, I set about understanding my own power consumption. All of it. It turns out that this is a tremendously difficult project, even if you have a rather large resource of tools and information available to you.

The easy pieces are those that are de facto measured for you (my values in brackets):

Plane travel miles [6,375 watts, 110,000 miles]
I can check all my plane tickets for a given year, use an assumption of plane fuel consumption (like 1.4MJ/km), and average that energy over the year to get a measure of the amount of power I'm using all the time.

Car travel miles [1,491 watts, 10,000 miles]
I can keep track of my odometers in the various cars that I drive and get an accurate number of miles, and by tracking fuel consumption over a number of tankfuls, I have an accurate estimate.

Home electricity use [135 watts]
My home electricity use is conveniently reported to me by Pacific Gas & Electric in the form of my bill, so I have accurate year-round data for this.

Home gas use [597 watts]
PG&E also gives me my gas consumption in a bill. I can convert from therms to watts.

I fly a lot. I don't drive much — when I do, it's mostly in a hybrid. I have a small house. Your values will probably be different, but not grossly so. These simple numbers are not, in fact, perfectly simple.

Do I count the air miles against me personally or my company? Do I count half of the car miles because I generally travel with someone else? Are the house utilities equally shared by me and my partner, or does she get more because of her penchant for turning up the heat?

Where it gets really difficult, however, is in the objects that I consume. My computer took energy to make, as did my clothes and my bicycles and my DVD cases and furniture. But no simple calculations for those yield a satisfying answer. So I looked in detail at the energy consumption of a simple drink.

For irony's sake, I chose an energy drink, but the calculation is reasonably applicable to any plastic bottled beverage. As I looked at the numbers, it was interesting to me that the drink already had an energy label, shown here, described as "nutrition facts." This energy label is for the chemical energy contained in the bottle that my body can convert to mechanical energy to do work, like riding my bike to the office. So I took the liberty of calculating the energy footprint of that bottle, in reasonable detail, and I summarized the results in an alternative bottle label called "consumption facts."

Let's think about how one would go about estimating the energy consumption of one bottle.

1. There's the energy embodied in the materials.
Embodied energy is a measure of the energy required to extract, purify, mix, and make the raw material. It is well calculated for many products. I'd have high confidence in this number.

2. There's the energy consumed in the transportation of the object to us.
The energy consumed in transportation is a little more difficult. How many miles did it travel? How efficient was the truck/train/airplane used to transport it? For this example, I assumed a 200-mile travel distance and an average 8mpg truck fully loaded with full energy drink bottles.

3. There's the energy used in manufacturing it.
The energy used in manufacturing is a little tricky for me to calculate. I'd like to divide the number of

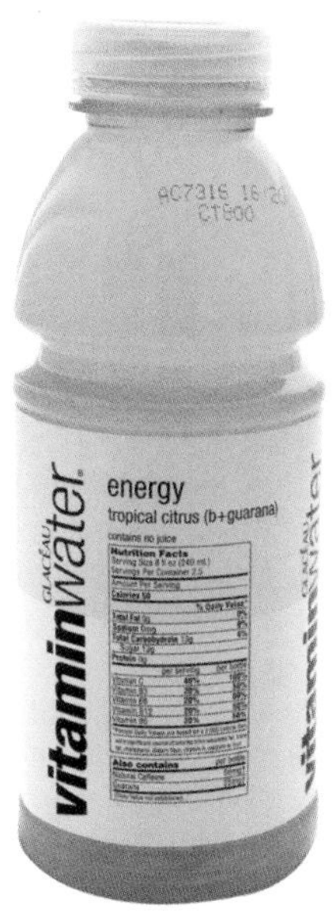

GLACÉAU
vitaminwa
contains no juice
Consumption Facts
Container Mass 1.58 oz (44.9 g)
Components Per Container 3
Embodied Energy Per Container
Total 4,609,420 Joules
PETE 38.81g 3,962,400 Joules
HDPE 4.83g 497,500 Joules
Cellulosic 1.34g 149,520 Joules
Recycle rate 23%
Landfill rate 43%
Energy recovery rate 16%
Lost to environmental waste 18%
Personal Energy Footprint % Daily Value*
Total 4.54%
Transport (avg.estimated) 0.69%
Manufacture 0.46%
Embodied Energy 2.67%
Refrigeration (avg.estimated) 0.71%
*Personal Energy Footprint is based on a recommended 2000Watt lifestyle.
The average US consumer has a 11400 Watt lifestyle.
! Consuming this product daily is equivalent to increasing your energy footprint by 90 Watts.
Also contains per bottle
Plasticizers 43mg†
Estrogen 0.12mg†
Carcinogenic Dye 0.19mg†
† Safe daily values not yet established

Photography and digital manipulation by Andrea Dunlap and Dan Benoit

empty bottles emerging from the bottle factory by the energy bill of the bottle factory, but not surprisingly no one really advertises that fact. The value I used here is a bit of a wild guess, but not off by more than a factor of 10, I'd say. It is a small value compared to the other values.

4. In the case of food, there's the energy used in its refrigeration and storage.
Finally, for the refrigeration of this product I took a shot in the dark. In the real world, this value should be a measure of the number of times it's cooled, warmed, and recooled, as well as the efficiency of the refrigerators, and the number of days spent on the refrigerator shelf. Again, I used a guess-like value, for illustration.

Those are the easy things to calculate. Harder things to add to the calculation would include the energy to run the computers of the workers in the company that produced the bottle, or the energy for the fluorescent lights in the retailer's shop.

Harder still are the negative effects of the non-energy, noneconomic factors of the product, like the aesthetic pollution they cause if washed up on beaches, or the health effects of toxic components like the plasticizers. This ethical calculus is very poorly developed, and in my mind one of the most important, difficult, and interesting areas of study in our times. If you're a student of economics, ethics, or philosophy, this is the frontier.

When I think about the future, it seems that buried in the maker ethos is a fundamental part of the solution.

What comes out of the analysis? Given my confidence in the embodied energy value, I can definitely say that a lower bound for the individual bottle of 5 million joules (MJ) and an upper bound estimate of 8MJ are reasonable.

That's the energy value; so how did I use this in the calculation of my daily allowance? Humanity currently uses around 15 terawatts (TW) of power. Watts are a rate, that is, an amount of power being consumed all the time: 100 joules consumed each second equals 100 watts.

Undoubtably, 15TW is a lot of power. It's hard to get an accurate figure, but to level off our CO_2 emissions to prevent global warming, we probably can only afford to make 2TW of that power with carbon-based fuels. The other 13TW would need to come from nuclear power and renewable sources that collectively produce barely 1.5TW right now.

There are 6.65 billion people on the planet. If you take the bold assumption that they all deserve an equal measure of the Earth's power resources, that would be 15×1012 / 6.65×109 = 2,255 watts each. The average American already uses more than 10,000 watts. I'm going to use 2,000 watts as your average daily allowance for two reasons: 1) It sounds fair. 2) It sounds familiar; we should consume a roughly 2,000-calorie diet to be healthy.

There are 60×60×24 = 86,400 seconds in a day. That means on the day you drink this bottle of energy drink, you are using about 90W. That's right, drinking that single bottle and disposing of it is like keeping an old incandescent light bulb on all day. If you drink a bottle every day, it's like you have your own personal light bulb following you around — always. Illuminating, eh? And I didn't even count the energy required to make the drink inside!

So why didn't I use carbon footprint as a measure? Well, CO_2 is sort of the enemy here, but it's the secondary effect caused by our lifestyles. If you calculate in terms of CO_2, it's tempting to believe you can "offset" that carbon by planting a tree. That might be true in the far future when carbon programs are well-managed and stable, but there's no guarantee that the tree you plant today will last forever and not simply return that CO_2 to the atmosphere when it dies. I prefer to measure our impact in SI units (International System of Units), because there can be nothing ambiguous about the results.

Every act of consumption we engage in has consequences — consequences that are not very visible to us. We aren't presented with the ramifications when we purchase the product, we don't see the direct repercussions of the single act, and our only way of seeing the end result is by abstract association with headlines like "Polar Bears Dying for Lack of Ice," or "Dolphins Choking on Plastic in the Pacific." I suspect we'd act a little better if we knew the consequences of our actions at the point of purchase. Here's an attempt to kick us off in that direction. Perhaps one day all products will sport both labels: nutrition facts and consumption facts.

Why would I want to bring this up in MAKE magazine? Partly because the solutions we seek are, in essence, engineering problems, and we need you to go out and start solving those problems. But also because when I think about the future, it doesn't look so good unless we change the way we generate and consume power. This will require everyone to change the way they do pretty much everything — and that's the good option. The bad scenario is that we don't change much, we kill all the supporting ecosystems, and descend into a future that makes *Tank Girl* look like a fairyland paradise.

And finally, because it seems to me that buried in the maker ethos is a fundamental part of the solution. Makers reuse things. Makers repurpose things. Makers repair things. I will bet that a handmade table built to last 200 years has an energy label not much different from an IKEA table built to last 7 years — except for the fact that you can amortize that energy over a time period almost 30 times as long. That means it effectively uses 1/30th of the energy.

High-quality things that are appreciated, repaired, and handmade become an important part of your life. My hope for a more beautiful future is that we will have fewer things pass through our lives, of higher quality, and love them more.

So go energize yourself by making an heirloom-quality, reusable water jug.

+ If you're interested in learning more, I recently gave an extended talk on this subject at O'Reilly's ETech conference, and the slides and accompanying text are available at wattzon.com.

Saul Griffith is a co-author of Howtoons and a recent recipient of a MacArthur Fellowship.

Work, Energy, Power

Work is the exertion of a force over some distance. I perform work on an apple when I lift it from the ground to a table.

Energy is the ability to do work. It's a measure of how much work you can do, whether it be moving apples, or heating your house.

Power is the rate at which you consume energy or do work. Lifting the apple quickly onto the table requires more power than doing it slowly, but the same amount of work is performed.

I chose watts as a convenient unit to do all my calculations in. Wattage is a measure of power, which makes it independent of time. People often ask "Watts per what?" The correct answer would be "Watts per always." It's the average. If you are burning a 100W light bulb, it's using 100 watts (joules per second) whenever it's turned on.

So I can conveniently use watts to add together the things I do that happen on markedly different time scales: the yearly things, the monthly things, and the daily things.

Thinking of your life in light bulbs might help you build an intuition for your power consumption.

Remember: "Watts per always."

Timothy and the Chocolate Factory

How a space shuttle technologist and the founder of *Wired* magazine hacked together a homebrew chocolate lab.

By David Pescovitz
Photography by Doug Adesko

VISITING TCHO IS THE MAKER'S EQUIVALENT of winning one of Willy Wonka's Golden Tickets. The new chocolate company, headquartered in a vast warehouse on San Francisco's Pier 17, is a sweet blend of DIY ingenuity, self-taught science, and online community. From hacking together a homebrew chocolate lab for $5,000 instead of buying a $100,000 "pro" system, to tricking out a 30-year-old chocolate factory line shipped over from Germany, TCHO (tcho.com) embodies a maker mindset that its founder calls "scrappy not crappy."

"I was a technologist working on things like the space shuttle when I got seduced by this weird, crystalline, alien goo that's called chocolate," says TCHO co-founder Timothy Childs.

TCHO, pronounced "cho," bills itself as a company "where technology meets chocolate, where Silicon Valley meets San Francisco food culture." Appropriately enough, then, the chocolate is still in beta. Fifty-gram bars in plain brown wrappers are available for $5 with formulations subject to change as often as every few days, incorporating direct feedback from, er, users. In the next year or so, though, the company plans to transform part of its warehouse into a retail store and European-style tasting room.

Childs' eventual goal is to see TCHO chocolate become an in-demand ingredient in other companies' products. His business plan is based on using the web to transform the supply chain into a supply loop. TCHO will use digital video and other media to tell the chocolate's life story, opening the lines of communication between, say, the Peruvian farmer who grew a particular bean and the customer on another continent. The entire manufacturing process will be transparent, he explains, from cacao pod to palette, and TCHO will be the hub of communication between supplier and sweet tooth.

The first step, though, is getting the factory line up and running. Matthew Heckert, a kinetic artist and alum of machine performance group Survival Research Laboratories, is leading the restoration of vats, tanks, mixers, and refiners that have been dry for years. The story of how the 1972-vintage factory line made its way to San Francisco is part of the whole TCHO creation myth.

In 2003, Childs, a veteran technologist, was working on a NASA contract to develop machine vision technology for the space shuttle. Around the same time, a friend at a previous tech startup had prophesied to Childs that premium, single-origin dark chocolate was the next big thing for foodies. Through a bit of kitchen chemistry, the friend had engineered dark chocolate to melt in your mouth. Childs was intrigued by the business possibilities.

Then, with the explosion of the shuttle Columbia on February 1, he suddenly had a lot more free time. Based on the new technique, Childs and his friend co-founded a confection company called Cabaret Chocolates. For production, they secured an old Oakland chocolate factory filled with obsolete, broken machines. During the day Childs repaired the machines, and at night he evangelized Cabaret at dot-com parties.

"I learned about how chocolate was made by fixing those machines," he says. "We had very limited investment, so we had to invent our own centrifuges

HOMEBREW CHOCOLATE: Timothy Childs (right) and Matt Heckert on the factory floor. Behind them is a conch and McIntyre refiner. Roxie, at lower right, is Timothy's dog and the company mascot. And no, Roxie doesn't risk her life sampling the product.

"We're virtualizing the whole control system so it can be operated from a single touchscreen. Eventually, I'll be able to run the factory from my iPhone."

and texture enhancement systems."

Childs' idea was to launch a line of premium chocolate to complement Cabaret's confections. When he realized the company wasn't moving in that direction, he set out alone. For months he immersed himself in chocolate, eventually finding himself at Interpak, the process and packaging machine fair in Germany where the candy industry converges every three years. There he met a "grizzled old chocolate expert," as he describes his adviser and eventual business partner (who prefers not to be named).

Shortly thereafter, this expert called Childs with news of a bargain. He was appraising a decommissioned East German chocolate factory that was selling off its top-notch but dated gear. Childs had to act quickly. He rang his old friend Louis Rossetto, the founder of *Wired* magazine and a Cabaret fan, and asked for a loan. Oddly enough, it was a perfect fit.

"Chocolate is a traditional food, but it's also very modern," says Rossetto, now the company's CEO and creative director. "TCHO isn't just selling chocolate but rather the whole chocolate experience. People want their products to tell stories and chocolate is the media for this story."

Once Heckert's team completes some repairs on the equipment, the line will be outfitted with the latest in digital sensors and video cameras. The sensors measure temperatures, monitor valves, and check for blockages while the cameras allow fewer people to operate the factory. The factory's original bureau-sized control panel, something that might belong on the Starship *Enterprise*'s bridge, is now a conversation piece in TCHO's reception area.

"We're virtualizing the whole control system so it can be operated from a single touchscreen," Childs says. "Eventually, I'll be able to run the factory from my iPhone."

In the meantime, Childs is gearing up the business operations and sampling ingredients. Not a bad job, except when the future of your business is hanging on your taste buds. Childs is like a sommelier of confections. In fact, his passion for wine predated his fascination with chocolate and directly inspired how he and Rossetto think about TCHO's product. "Timothy developed a new taxonomy for chocolate," Rossetto says.

Today, most premium chocolate is classified by region or by percentage, referring to the mass of cacao bean (cocoa) in the bar compared to flavorings like sugar or vanilla. The problem, Childs explains, is that percentages aren't a true gauge of flavor. Different beans can be coaxed to yield various flavors by adding more or less sugar or tweaking the manufacturing process, so two bars may have similar percentages of cocoa mass in them, but taste quite different. Classifying by origin is equally problematic because most regions offer many different flavors of beans.

"Wine has similar descriptors as chocolate," Childs says. "So I decided to use six base-level, building-block flavors to classify our chocolate, like fruity, nutty, and, of course, chocolatey." The TCHO Flavor Profile acts as a road map to define the types of beans he seeks in South America and elsewhere. Once Childs connects with a farmer, he heads back to the lab to put some of their beans through their paces (*see "From Pod to Palette," page 32*).

TCHO is slowly outfitting the chocolate lab with industry-standard equipment, but the company was launched with scavenged gear hacked together with duct tape and glue. It wasn't pretty, but it worked. And it served as a prototype. TCHO's plan is to help farmers build their own similar labs.

"Most independent farmers have never tasted chocolate from their own beans," Childs says. Providing farmers with tools to improve their product, he explains, will help them shift from producers of commodity beans to premium growers. And that can only lead to better chocolate for everyone.

"TCHO," he says, "is really a tool to reshape how premium chocolate is made and experienced."

David Pescovitz is editor at large of MAKE.

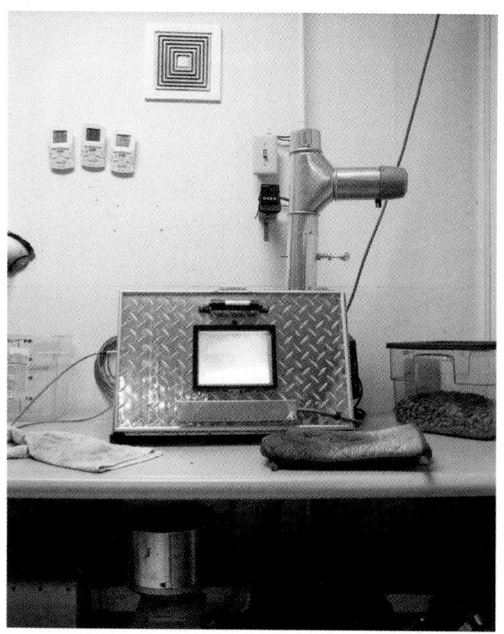

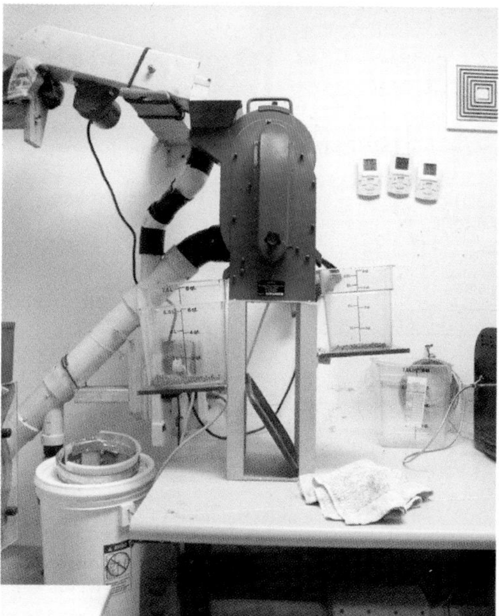

TOP LEFT: A converted and customized coffee bean roaster works like a champ for cacao. TOP RIGHT: The Crankenstein is a grinder for beer hops, customized with a wider gap between the rollers for cocoa beans to fit through. BOTTOM LEFT: After it's ground, winnowers are used to blow off the bean's shell from the meat. Childs modified plans he found online for a winnower designed for agricultural use in developing nations. They cobbled together a spare blower from Heckert's shop using a Shop-Vac found on a sidewalk, and made a vibrating feeder using a vibrator from another old machine. After adding some PVC pipe, the total cost was $12.

RIGHT: These wet mixers, now used as chocolate melangers, were made to mix Indian curries. Temperature control is essential when making chocolate, so Heckert took Childs' design and outfitted the mixers with infrared sensors and cheap space heaters that blow just the right amount of hot air on the oozing goodness.

CLOCKWISE FROM TOP LEFT: In the hot room, chocolate is poured and molded by hand. Adjacent to the hot room is the cool room, where the bars are demolded. Stackable bins hold a variety of chocolate samples that are bagged and labeled by batch.

For an online photo gallery of TCHO's chocolate factory, see makezine.com/14/proto.

FROM POD TO PALETTE

By Timothy Childs, founder and chief chocolate officer of TCHO

Chocolate starts from cacao beans, which grow inside strange looking, geometrically complex pods that sprout directly out of the limbs and trunks of cacao trees in the tropics.

There are four main phases for transforming cacao beans into finished chocolate, and each phase generally has 3–4 steps:

1. Growing: planting, harvesting, fermentation, and drying.

After the cocoa pods are harvested, the beans are placed in boxes or heaps where they undergo active fermentation for 5–7 days. Fermentation cooks the beans in their own juices to around 50°C (122°F).

Proper fermentation first stops the bean's growth, and then chemically alters its composition. It's this chemical transformation that coaxes out the more complex and delicious flavors of cocoa beans. With proper fermentation, you can make a good bean great, and a great bean spectacular. While most beans processed in the world are fermented poorly or not at all, I only use very well fermented beans.

2. Processing: cleaning, roasting, cracking, and winnowing.

Once the bean has been harvested, fermented, dried, and roasted, we crack open the bean to separate the shell from the meat of the bean (called nibs) in a process called winnowing. When heated and crushed, the nibs turn to a molten state, called cocoa liquor.

3. Refining: mixing, refining, and conching.

Sugar, additional cocoa butter, and vanilla are mixed in with the liquor, then this mixture is refined down to sub-20-micron size. Next, the almost-finished chocolate is conched (a process that blows off excess acids to improve flavor) for up to 72 hours.

4. Molding: tempering, molding, cooling, and wrapping.

The finished chocolate is then tempered (a process that forms crystals that help it solidify nicely when cooled) on its way to being squirted into molds for making bars. Yum!

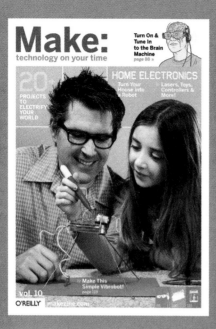

NO POSTAGE
NECESSARY
IF MAILED
IN THE
UNITED STATES

BUSINESS REPLY MAIL
FIRST-CLASS MAIL PERMIT NO 865 NORTH HOLLYWOOD CA

POSTAGE WILL BE PAID BY ADDRESSEE

Make:
PO BOX 17046
NORTH HOLLYWOOD CA 91615-9588

NO POSTAGE
NECESSARY
IF MAILED
IN THE
UNITED STATES

BUSINESS REPLY MAIL
FIRST-CLASS MAIL PERMIT NO 865 NORTH HOLLYWOOD CA

POSTAGE WILL BE PAID BY ADDRESSEE

PO BOX 17046
NORTH HOLLYWOOD CA 91615-9588

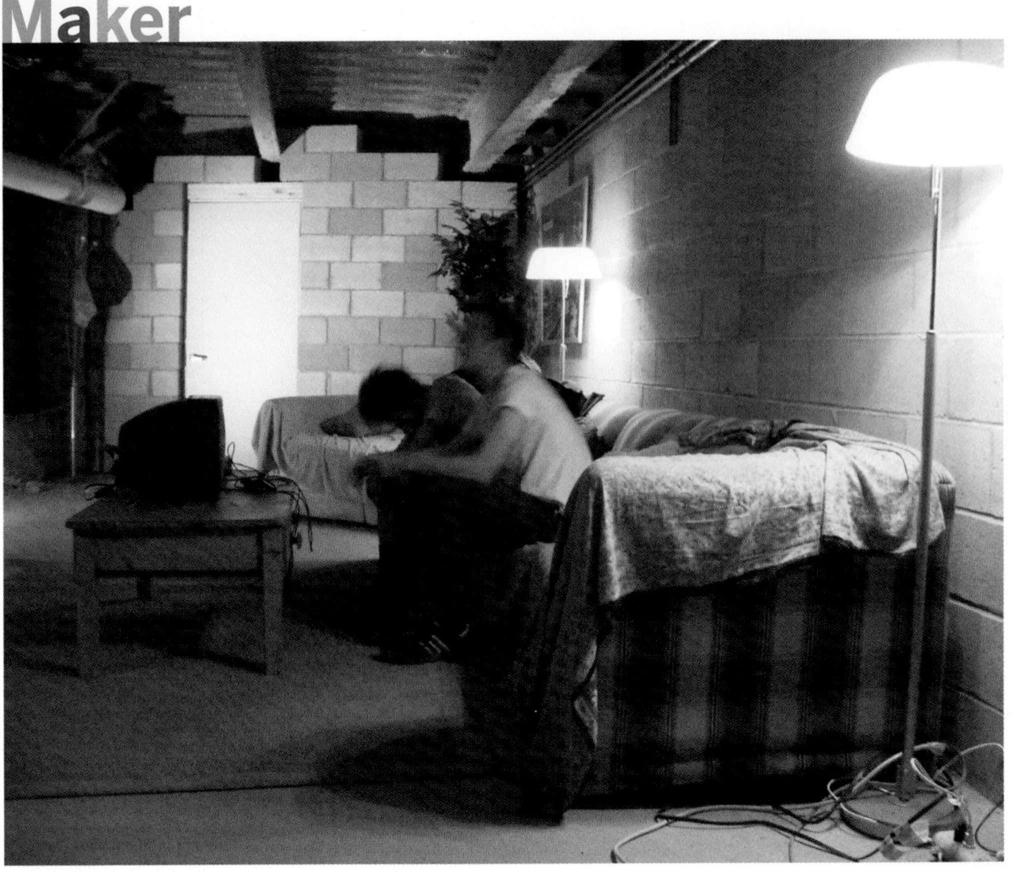

Mall Living

Artist Michael Townsend and his wife, Adriana Yoto, lived in a 750-square-foot space, rent-free, for four years. The Providence, R.I., couple managed to build an apartment inside the parking garage of a mall, all the while evading mall personnel and security. By Howard Wen

Thirty-six-year-old Michael Townsend and Adriana Yoto, 29, unofficially (and illegally) established a residence within Providence Place, a behemoth shopping and entertainment complex. In 2003, they took over an empty 750-square-foot storage room, which was walled off on three sides, in the mall's parking garage.

With the help of a couple of friends, they snuck in 90 cinder blocks and an industrial door to build a fourth wall to close off the area. Inside, they set up the usual apartment furnishings: couches, a rug, a coffee table, lamps, a TV set, a china hutch, and even portraits on the walls. Whenever possible, they bought these items from stores in the mall itself. They ran electricity from an electrical cable connected to a working outlet in the parking garage. In all, Townsend estimates that they spent about $5,000.

While they weren't homeless (the couple also lived in their own separate studio apartment), Townsend and Yoto and their friends spent quite a lot of time hanging out in their mall homestead.

Photograph by Michael Trummerkind

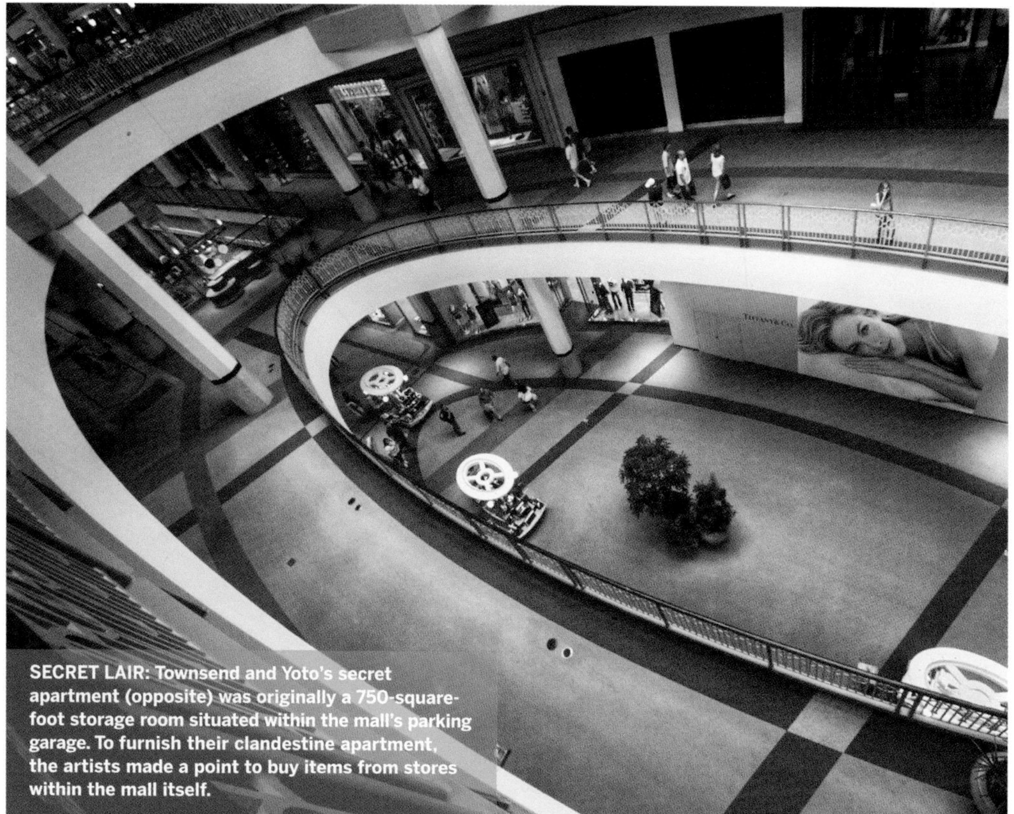

SECRET LAIR: Townsend and Yoto's secret apartment (opposite) was originally a 750-square-foot storage room situated within the mall's parking garage. To furnish their clandestine apartment, the artists made a point to buy items from stores within the mall itself.

Photograph by Jason Liu

And things went on this way for four years — until late September 2007, when Townsend was caught by mall security and arrested by the police. He was charged with a misdemeanor for trespassing and given six months probation.

Howard Wen: What was appealing about living at the mall?

Adriana Yoto: We wanted to create a space that embodied the design ideology of the mall. Our end goal was to create a space that one would think was actually part of the mall. We wanted people to feel like they were walking into Crate & Barrel or into a page of *Domino* magazine.

HW: The interior design of a lot of mall stores is indeed appealing. So that's what you were going for in your space — re-creating that kind of design sensibility?

AY: The micromanagement of creating clutter-free homes and all these different storage techniques, creating the perfect pantry, the perfect closets — I think a critical culture has emerged recently of hyperorganized homes prepared to be seen by guests at any moment.

HW: Before you got caught, did you have an end game in mind with this project? Was there ever a point planned in the future when you would be finished with it?

AY: We never thought of it as a "project" — it was sort of this hobby integrated into our lives, and we never thought anybody would find out about it. We never intended it to be public. So we never imagined the end. We just imagined constantly improving and evolving it.

HW: How did you sneak into the apartment each day?

Michael Townsend: There was an architectural anomaly in the building that essentially acted as a private entranceway for us. It was a passage between two walls that was at its narrowest about 11 inches and spread to about 2 feet. It was about 60 feet deep. If you stood outside the mall, in the right place, and threw a baseball with accuracy, you could throw it into our home. Once you went

WITHIN THESE WALLS: Providence Place (red brick building to the right) is a gigantic shopping, dining, and entertainment complex in Providence, R.I.

through this, and were in the space, you were then connected to exit doors and the system of doors and stairways that provided exits for the mall.

HW: Were there limited times that you could go in and out?

MT: No, we had 24-hour access through our private entranceway. And the garage is open 24 hours a day too.

HW: Did you have to walk through a parking garage or something like that?

MT: There were entrances from the garage portion of the mall to our space. The garage is architecturally a little over half of the mall.

HW: Were there security guards you had to go by? Security cameras?

MT: There is 24-hour security and a standard network of cameras. As it should be, it is not an overwhelming or oppressive amount, so movement is not restrictive. In the garage or through the mall, there are precautions to take, but through our private entranceway it was a clean shot 24/7.

HW: How did you get the building materials and furniture in?

MT: If it was really thin, like lamps, doors, cinder blocks, rugs, clothes, artwork — those could go through the "squeeze hole," as we called it.

If it was larger, we had to be very, very, very careful and just get the object through any door that led into a network of stairs and hallways that was ultimately attached to our nondescript door entrance (which we could open from the other side or was left ajar most of the time.) Once you got through any door and were not visible, you were pretty safe.

HW: How did you shower and use the toilet?

MT: Bathrooms were in the mall. If we had been allowed to be there for the [full] year, we were going to get a membership at the Westin gym. The Westin is a luxury apartment/hotel attached to the mall, and we would have worked out there

Photograph by Richard Benjamin; illustration based on Providence Journal map by George Silvia

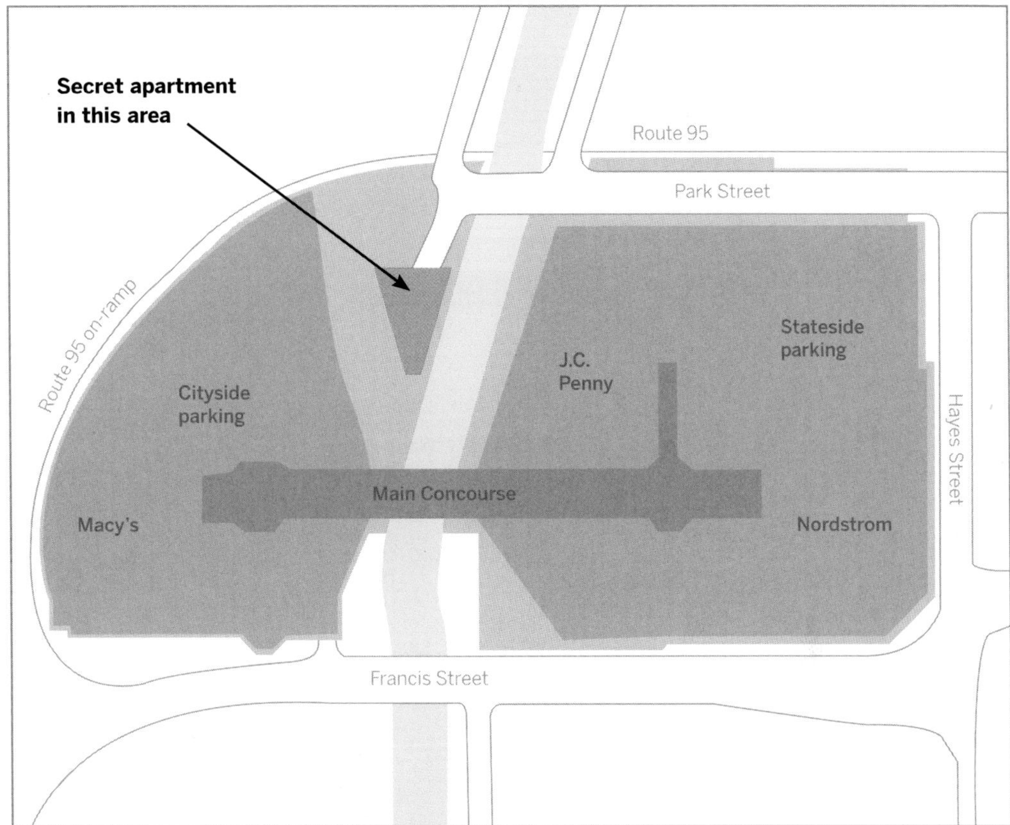

in the morning and then used their shower. But, alas, our dreams of getting buff and staying clean were shattered.

HW: Michael, you have come across in the news media as extraordinarily gracious toward the police. Were you surprised by the way they treated you, especially in this day and age of heightened security concerns?

MT: We were in a high state of panic nearly all the time. But we knew if the space reached a "plateau of completion," the actions against us would be lessened. It's an odd calculation.

Before the judge, they started reading the charges against me. They said, "He entered a storage closet, which gave him access to a ladder that went up to a loft space, where over the years he created an apartment."

I'm watching the judge, and I see [his] expression change, and they start describing the apartment in full detail: "The apartment was fully furnished, had sectional couches, matching lamps ...." At that point, the judge called everyone close to him, and they talked for a while and said, "We gotta give this kid a misdemeanor charge."

The cop who brought me to the police station said, "How am I going to write this up?" I was like: "Well, trespassing with an intent to decorate?"

AY: I think our commitment was our redemption.

HW: Do you have anything to tell others who are entertaining the idea of doing what you guys did? Or a warning?

MT: It's worth noting that you can put a cinder block in a backpack. That means you can build a house anywhere you want.

+ Townsend and Yoto's "Living in the Mall" project: trummerkind.com

---

Howard Wen reports on technology and tech culture for several publications and websites. He can be reached at howardwen.com.

Cars Without Drivers

Teams compete to win the U.S. military's $2 million grand prize.

By William Gurstelle

The Pentagon's Defense Advanced Research Projects Agency is famous for its pursuit of high-risk, high-reward technologies. When DARPA bets on technology, the wins or losses can be spectacular. Its latest big bet came in the form of the third edition of the DARPA Grand Challenge.

Known as the "Urban Challenge," the contest took place on a sunny November weekend at DARPAtown, a specially setup racecourse on the grounds of the vacant George Air Force Base in Victorville, Calif.

There, some of the best and brightest makers in the country were lured into tackling a difficult problem that strongly appeals to their love of making things. Unlike DARPA's typical top-secret projects, the Urban Challenge was designed to publicly showcase the talents of top-level makers. And the $2 million top prize doesn't detract from the appeal either.

DARPA's war planners want a way to keep soldiers safe during dangerous supply missions. Their goal is to develop a vehicle that can drive itself to a dangerous place and do what it needs to do — drop off gear, deliver supplies, or pick up soldiers — without risking the life of a human driver.

Toward this end, DARPA organized the Urban Challenge. The barriers to entering the contest are relatively low — any qualified team of engineers and mechanics can compete. But winning the challenge is a task of Herculean dimensions. To win, the team must build a car that can autonomously maneuver a 60-mile course in an urban environment, "executing simulated military supply missions while merging into moving traffic, navigating traffic circles, negotiating busy intersections, and avoiding obstacles."

Photography by William Gurstelle

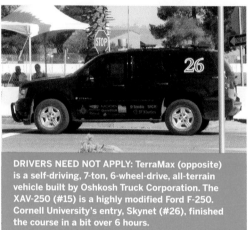

DRIVERS NEED NOT APPLY: TerraMax (opposite) is a self-driving, 7-ton, 6-wheel-drive, all-terrain vehicle built by Oshkosh Truck Corporation. The XAV-250 (#15) is a highly modified Ford F-250. Cornell University's entry, Skynet (#26), finished the course in a bit over 6 hours.

Several rounds of tough competition eliminated most of the contestants prior to the competition's final day. Just 11 vehicles out of the 36 semifinalists were still in the hunt. Those remaining were given three missions to complete. Moments before race time, the teams were provided with details of their secret missions, the information provided as a computer file on a USB jump drive handed to each team's leader. Each mission was different, requiring the vehicle to negotiate through the sometimes heavy DARPAtown traffic. To be eligible to win, a vehicle had to successfully complete all three missions in less than six hours.

At first look, designing a self-driving car may seem impossible. But really, "it's just an extension of present technology," says Michael Darms, an engineer with the Tartan Racing team. "Cruise control, which is a first step, has been around for decades."

Darms lists numerous examples of existing cars that handle more and more of the tasks of driving. Many luxury cars have smart cruise control that automatically maintains a safe distance from the cars ahead. Further, some have anticipatory braking systems that use radar to anticipate a crash and pre-charge the brakes for a faster stop.

The vehicles at Urban Challenge must go beyond simply removing the driver's foot. They must remove the driver's brain.

But the vehicles at Urban Challenge must go beyond simply removing the driver's foot. They must remove the driver's brain. Doing so requires a great deal of technology. In use here are radar, LIDAR (light detection and ranging), gyroscopes, and machine vision sensors, all of which paint an incredibly detailed digital image of the area surrounding the vehicle for use by the onboard computer. Control schemes include sophisticated "fly-by-wire" systems that turn the car's wheels and apply its brakes. And the GPS systems used are incredibly accurate, telling the car where it's located to within a few centimeters.

The teams at the Urban Challenge range from huge groups from corporations and universities with millions of dollars in funding behind them, to groups of five or six tinkerers who modified their personal cars. As one might expect, the greater the resources, the better the self-driving cars perform, for money does matter. But all of the robot cars here, even the low-budget ones, perform admirably. Here are some typical Urban Challenge entries, ranging from the simplest to the most elaborate.

SIMPLEST: ODY-ERA

While other entrants have paint jobs proclaiming Ford, Caterpillar, and Google as sponsors, Ody-Era's decals include Papa's Italian Bistro and Mac's Fabrication Shop. Ody-Era, a 2008 Mercury Mariner from Kokomo, Ind., may not be the most sophisticated, but the fact is, its makers legitimately qualified to compete at DARPAtown against teams a thousand times larger and richer. They showed that a few makers working in a home garage can still attempt great things.

"We've spent less than $20,000 in total on our vehicle," says Rick Bletsis, the driving force behind this self-driving car. Yet Ody-Era made it through several levels of competition in order to compete in the Urban Challenge at DARPAtown.

Unlike most other competitors, Ody-Era relies mostly on machine vision to guide itself through the course. "We use inexpensive, off-the-shelf digital

LOW BUDGET: Mahesh Chengalva wrote all the code for the low-budget, high-ambition Ody-Era (#55). Other robot cars had so much computer power that they required onboard gen-sets for power and air conditioning.

cameras and a desktop computer to sense the environment. A local farmer lent us the John Deere GPS from his farm tractor," says team member and software engineer Mahesh Chengalva.

With simplicity as its watchword, the car utilizes a program consisting of only 4,000 lines of Visual Basic code to drive itself. Contrast that with the hundreds of thousands of lines of code inside the control algorithms of many competitors.

Ody-Era passed several key hurdles but was stopped well before the finals, undone by a problem with the computer that controlled its drive-by-wire steering.

The team has few regrets. "We did our best, that's all we can say," says Bletsis. "Our technology is so novel, nobody else here has much like it."

SIMPLER: PLAN B

If the first idea didn't work, then maybe it's time to consider Plan B. Many would find it surprising that two brothers who run an insurance company in New Orleans would have the interest and the technical know-how to develop an autonomous vehicle able to compete with bigger, richer teams like Tartan Racing and Stanford University. But the Gray brothers are serious about robot drivers, and they've developed their own autonomous vehicle called Plan B.

Plan B is a Ford Escape Hybrid packed with sensors and Linux-based computer intelligence. If your map is good, they say, you merely need to tell the vehicle where to go and let the SUV find its own way.

For just $365,000, give or take a few thousand, Team Gray Racing will sell you a real, live autonomous vehicle, complete with an Oxford Technical Solutions GPS system accurate to 10 centimeters, a Velodyne 3D high-definition LIDAR sensor, a fly-by-wire steering and braking system, and Gray Racing's own AVS-2 intelligent driving computer with obstacle detection and dynamic rerouting capabilities.

Team Gray Racing did very well in the 2005 DARPA Grand Challenge, being one of only five teams to finish. It would have been a miracle if little Team Gray had reached the finals in the 2007 Urban Challenge. But there was no miracle this time, for Plan B was axed in the final cut, the day before the final event.

NOT SIMPLE: BOSS

Boss, an autonomous 2007 Chevy Tahoe, is like the

THE FIRST ROBOTIC FENDER BENDER

I must admit, there is a certain amount of control I'd be happy to turn over to the car, if I really trusted it. But let it drive itself? I'm still a long away from that.

And to this I feel I can speak with authority, for I was the human closest to the first-ever crash of two fully robotic vehicles, one that took place just feet from where I stood observing the race.

The interesting thing was the incredible "human-ness" of the crash. It was as if both robots were imbued with stereotypically human behind-the-wheel personalities. MIT's Talos, described as the world's most expensive Land Rover, did indeed behave like an impatient, self-entitled luxury car, while Cornell's Skynet, a modified Chevy Tahoe, acted timidly and with maddening uncertainty, like an out-of-town tourist in an unfamiliar rental car.

Upon seeing the accident occur as I did, if you didn't know the cars had no drivers, you would think it was simply a typical case of poor driving, not a failure of computer technology. One car was too aggressive, becoming impatient when the other car couldn't decide what to do at an intersection. Talos decided to pass the hesitating Skynet and at that moment, the Tahoe hit the gas to move forward. Crrrunch! ***—William Gurstelle***

BIG MONEY, BIG PAYOFF: Stanley (#3), Stanford University's VW Passat, came in second, while Carnegie Mellon's Boss (#19) drove off with first place, finishing the course in a stunning 4 hours, 10 minutes.

For just $365,000 Team Gray Racing will sell you an autonomous vehicle, complete with a GPS system accurate to 10 centimeters.

New York Yankees: it's a huge monetary investment whose sponsors expect a big payoff. The Tartan Racing team, an amalgamation of academics, engineers, and students from Carnegie Mellon University, General Motors, and Continental AG among others, built Boss, outfitting it with a complex array of radar, laser sensors, and drive-by-wire control systems.

Technology like this doesn't come cheap, and a multimillion-dollar project like Boss requires the combined financial resources of several global conglomerates. Say what you will about design elegance and resourcefulness; in the end, money talks. That's why the Yankees win so often.

First place and the $2 million grand prize went to Tartan Racing's Boss, for completing all three missions 20 minutes ahead of the competition. Stanford University's Junior came home with a check for $1 million, and Virginia Tech's plucky Odin got half a million.

All of these teams were deep-pocketed and loaded with the best technical minds available anywhere. But big-thinking amateur makers shall not be dissuaded. As they say in New Orleans and Kokomo, wait until next time.

William Gurstelle is a contributing editor for MAKE and author of the popular book *Backyard Ballistics*.

Saucy Recycling
Don't forget to recycle interesting food containers that may have convenient use in the shop. This Worcestershire sauce bottle works perfectly for dispensing a slow stream of solvent. Be sure to remove the label and add the appropriate solvent label for use in the shop.

Find more tools-n-tips at makezine.com/tnt.

Flying School Automaton

Technology-loving artist Diane Landry discusses her background, her inspiration, and how kids respond to kinetic sculpture. By Annie Buckley

Walking into Diane Landry's installation *Ecole d'aviation* (*Flying School*) is like stumbling upon a magical garden. Dancing umbrellas take the place of flowers, the vibe is eternally moonlit, and a beautiful medley of mechanized harmonica parts fills the air like mystical birdsongs.

For the past 20 years, Landry has used various kinds of technology to transform everyday objects into her uniquely theatrical installation art. Her roots as a performance artist are playfully evident in her animated sculptural works that have been exhibited internationally since 1990. I had the chance to speak with her by phone from New York, where she's currently enjoying a six-month residency in the SoHo district of Manhattan as the winner of a Québec grant competition.

Annie Buckley: A bit of background first: where did you grow up? Did you make art or use technology as a child?

Photography courtesy of SolwayJones, Los Angeles (this page) and Diane Landry (opposite)

OPPOSITE: *Flying School* (2000) uses 24 umbrellas, harmonicas, motors, steel, halogen lamps, a MIDI controller, and a computer. ABOVE: *Privileges* (2006) features a flip book with automation.

Diane Landry: I grew up French Canadian in Québec City. For me, drawing was always part of my life, but a child just draws or paints. I wasn't thinking that this could be a job.

AB: I read that you previously worked for the Canadian government. When did you transition to being an artist?

DL: I received my degree in technical natural science and then found a job in the government doing research in agriculture. I really liked the work, but after five years, I just wanted to be an artist. I quit my job to go back to school in fine arts. That's when I found my place as an artist.

AB: *Flying School* was presented around the time of the Katrina disaster. How did that event change or alter the work?

DL: I made that work in 2000 and the project traveled quite a lot. In 2005, I presented it in Houston, just one month after Katrina, and during another hurricane. Most people left the city, but we decided to stay there and sleep in the gallery under the umbrellas. That was quite a nice experience.

I had decided to work with umbrellas because of their relationship to weather and to water, but we can read my work from different perspectives. It's about recycling objects, but also recycling the meaning of objects. I like to make projects that are open-ended and have many different readings.

> "Children are really open. They are like, 'Oh wow, it's a new friend.' They are not thinking, 'Is it OK for sculpture to move?'"

AB: What role does technology play in your process?

DL: Technology can create motion, or an idea of life. When you recycle objects, you give them a new purpose, but I also want to give them life in the sense of moving, like the motion of breathing. The best way to do this was to work with motors and sequences and all that technology. I started with basics, just adding light to installations, and I enjoyed it. I saw that changing the light could change the work, so early on, that was part of my process. To include motion and electricity was a natural next step for me.

AB: I notice that children often respond more favorably than adults to art that incorporates technology. Have you noticed that too?

DL: Definitely, children are really open. They are like, 'Oh wow, it's a new friend.' They are not thinking, 'Is it OK for sculpture to move?' They just enjoy it. Most of the time you see all of the technology in my work. There is no mystery — everything is there.

To see more of Landry's work in action, visit dianelandry.com and solwayjonesgallery.com.

Annie Buckley is a writer and artist in Los Angeles. Her collection of short stories, *Navigating Ghosts*, was published by Nothing Moments in 2007. anniebuckley.com

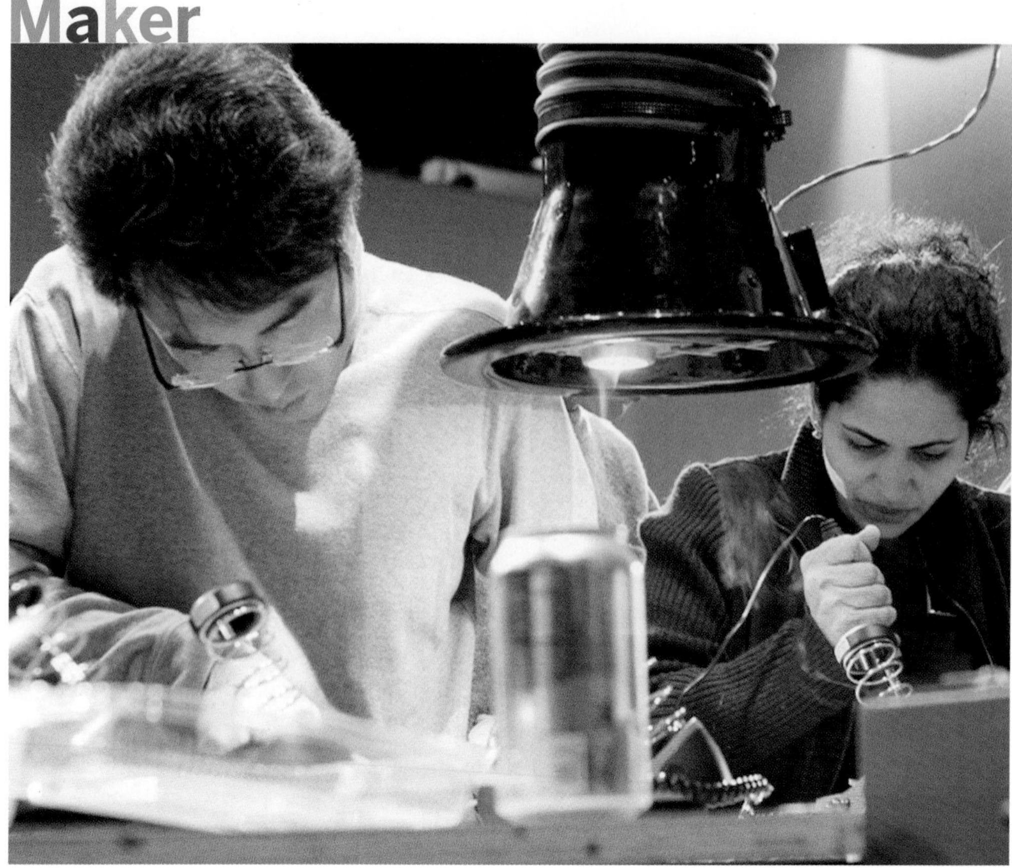

Ready, Set, Make!

Twenty-five teams, 1 box of parts, 4 hours, winner takes all — welcome to the 2008 XD Design Build Prize.

By Gever Tulley

Sometimes you come across an object, device, or tool so compelling that you make up excuses to use it. This is how I feel about the Arduino, an open source, low-cost, easy-to-use electronics prototyping system designed to be used by people of all skill levels, from rank beginner to high wizard (*see MAKE, Volume 07, page 52, "Arduino Fever"*).

That's why I hosted a contest at an internal Technology Summit held biennially by Adobe Systems, where I work as a senior computer scientist, in San Jose, Calif. The goal of the event was to put designers and developers on equally unfamiliar ground, while at the same time providing a problem to solve that would require the best from both mindsets. The loftier goal was to help them learn to appreciate each other, but we weren't holding our breath on that one. Robyn Orr and Julie Spiegler, frequent collaborators on my hairbrained schemes, agreed to help me make it happen.

Knowing that developers are less comfortable with a subjective measurement of success, we crafted a goal that could be empirically determined: "Create an object that will keep a judge engaged for the longest amount of time," with engagement* defined as "from the time the judge opens the box containing your object, to when they close the box."

In order to keep the judges from self-consciously

*Working for Adobe Systems (creator of the Engagement Platform), I was reminded that for this project we used the definition of *engage*, "to occupy the attention or efforts of a person," not "to attract or please."

Photography by Bob Murata

adjusting their engagement, they would not be told about the time factor and instead would be asked to give each object a completely subjective rating based on a scale of one to five stars.

Now we had the idea and a good working definition; all that remained was to work out a budget and parts list. The Arduino runs about $35, so we decided on a budget of $75, leaving $40 for parts. Some of that would be eaten up by nonnegotiable items like the USB cable necessary to program the Arduino, a 9-volt battery, a battery clip, the box the parts would all fit in, and tools each team would need to work with. Whatever was left had to sustain the creative needs of the designers within the technological capabilities of the system.

The folks at MAKE enthusiastically agreed to loan us workbenches and to supply some covetable MAKE items to supplement the grand prize.

With the challenge established, we refined the parts list. Since the idea started with the Arduino, that's where we started as well. We knew we needed some forms of packaging that would accommodate the circuit board, a breadboard, and at least a battery. None of the traditional electronics project boxes had the tactile aesthetics we wanted, so we looked elsewhere for suitable alternatives. We found them in the aisles of Ikea: plastic dishware and food storage containers in lots of colors, shapes, and sizes — all easy to hack with a drill and a box cutter.

Then there was a series of thought experiments about what kinds of experiences could be built into these containers. Holding the plastic cups and boxes in our hands, we said things like, "You could put switches here and here and a couple of lights to make a puzzle. What about a little R2-D2 thing that makes beeps and chirps? Wouldn't it be cool if ...?"

That got us to a list that we separated into input and output capabilities. With only four hours to learn to program the Arduino, build the object, and wire it up, we decided to eliminate the more exotic items that required sophisticated development and focused on collecting reliable and fault-tolerant components — switches, lights, speakers, and a little motor normally used to make cellphones buzz. For fun we added three-color LEDs that could, with a little extra effort, be made to shine in almost any color, and a couple of photosensors that would be easy to connect to the analog inputs of the Arduino.

The switches and some of the LEDs were bought in bulk 100-piece grab bags, so rather than risk giving one team better parts by accident, we put poker chips in each box that could be traded for parts from a common table. One chip bought exactly one switch, light, cup, straw, plate, or box. We were feeling generous, so teams could take a whole handful of the little plastic beads and assorted doohickeys.

We also allowed teams to barter with each other, or trade a part they didn't want for one they did (as long as it was available on the parts table). No intrinsic valuation of parts was made and a strict one-for-one basis made for some interesting trades: one toggle switch for one red straw, one possibly burned-out LED for one lime-green bowl. Many suggestions were ignored about establishing a currency that valued the multicolored LEDs higher than the invisible infrared LEDs. The uncomplimentary lighting in the room resulted in wholesale dumping of the photosensors, while the pressure switches were surprisingly popular.

To prepare for the event, we filled each box with

TOP: Senior Director Winston Hendrickson is perplexed by the stimulus/response mechanisms in the electric mushroom. BOTTOM: The plasticware proves tougher than expected and the supplied box cutters are abandoned for a turn at the soldering stations.

identical supplies, including a CD with some sample code and simple documentation. We arrayed the remaining items on the parts table, set up shared tools on the tools table, and powered up the projector to display the simple rules. We displayed the prizes and MAKE schwag prominently at the front of the room, and of course we had T-shirts printed up to commemorate the event.

I stood behind the podium and looked out over the empty MAKE tables. In just a few moments we would open the doors and let the teams in, and chaos would almost certainly ensue. There had been last-minute panics and near cancellation of the event when it was learned that the fire alarms might go off if we used soldering irons inside the auditorium, but with some deft negotiation and the addition of an industrial fume extraction system that looked like it was going to inhale some of the contestants, we got the green light.

Robyn opened the doors and as the teams spilled

The level of excitement in the room rose as strange objects took shape, blinking and buzzing to announce their arrival.

into the room, Julie used an LED and a resistor as a litmus test to determine technical savvy and ensure that each team had someone who remembered something about electronics ("Do you know what these are?").

The time passed all too quickly. At the one-hour mark, few of the teams had the Arduino's "Hello World" blinking light working. Many were still making drawings on notepads and holding cups and plates to illustrate how their device might work. At two hours, only a few teams hadn't yet got the light to blink, and furious trading and general construction and development had begun. The level of excitement in the room rose as strange objects took shape, blinking and buzzing to announce their arrival. At three hours, general activity rose to a frenzied pitch.

Original plans were abandoned as the deadline loomed, and hasty fallback solutions were crafted. The last team not to get a blinking light suddenly realized that their Arduino was actually broken and it was swapped for a spare.

At the announcement of "Fifteen minutes to go!" moans and groans erupted among the contestants, and emissaries were dispatched to the podium to negotiate for more time. With five minutes left, some teams were trying to dress up their creations and pass off a simple blinking light as a game or toy, but two of the teams were completely done, leisurely spending the last few moments to review their code and weed out any last bugs. At "Time's up!" most of the teams began reluctantly scooping partially completed devices into the judging boxes. Under the strain of being lifted, wires pulled free and some of the devices did not survive the boxification process.

Judging took place in a conference room ten floors up, where we'd gathered some of the digerati of Adobe to evaluate the projects. If getting the objects into the box was a risky endeavor, getting them out proved to be downright dangerous, and many more finished objects perished at the judges' tables. However, the nonworking objects proved to be as fascinating as the working ones, and by our strict rules, many of the broken ones scored as high or higher than the working objects. Of the working ones, a polyphonic flute/trumpet easily proved to be the judges' highest subjectively ranked object.

The day's clear winner, based on total time of engagement, was a nonworking puzzle game that involved dropping colored beads down channels that led past photosensors. Tantalizingly, a blinking light deep inside the object suggested that if beads were dropped at the right time, something would happen. But nothing did happen, and the judges gave up playing with it, but only after many minutes of trying.

Two days later prizes were awarded at a simple ceremony, and the excitement over the grand prize paled in comparison to the absolute glee expressed by the teams when they were given back their creations. Later, we found some of the teams gathered around tables resurrecting their devices.

The overwhelming response from participants has been, "Let's do it again, only we get all day to do it — I know exactly what we're going to make next time!" Well, we're changing the rules next time, and there's going to be a new list of parts to contend with....

Gever Tulley (gever@tinkeringschool.com) is founder of the Tinkering School, senior computer scientist at Adobe Systems, and general fool-around guy.

Pushing the Performance Envelope

Mary Hallock-Greenewalt was a musician, engineer, inventor ... and exceptional.

By Michael Betancourt

Photograph courtesy of The Historical Society of Pennsylvania

Even today, Mary Hallock-Greenewalt (1871–1950) would be unusual, but in the early years of the 20th century, she was exceptional. Between 1919 and 1927 she filed 11 patents, while performing as a piano soloist with both the Philadelphia and Pittsburgh orchestras, recording with Columbia Records, and producing hand-colored "films" along with a machine that timed them to play in sync with music.

Hallock-Greenewalt's various patents described the components for a new, technological art form she called Nourathar, derived from the Arabic for "essence of light." Her instrument for performing this light music, the Sarabet, employed mercury switches, motor-controlled lights, and rheostat switches, aka dimmers. Her patents were so revolutionary that they were almost immediately stolen for use in theatrical lighting manufactured by General Electric. She sued GE for patent infringement, ultimately winning on appeal — the first judge didn't believe that as a woman, she had invented so complex an electrical device as the rheostat.

Hallock-Greenewalt's Sarabet, named in honor of her mother Sarah Beth, underwent continuous refinement and tinkering. She often changed the placement and number of lamps to achieve a full coverage of the performance space's geometry and architecture. The Sarabet controlled a network of 1,500-watt lights that could be dimmed individually and turned red, orange, yellow, green, blue, or violet by filters on color wheels. More complex hues were mixed on the performance screen.

In place of a keyboard, the Sarabet had a console with graduated sliders and other controls, more like a modern mixing board. Lights could be adjusted directly via the sliders, through use of a pedal, and with toggle switches that worked like individual keys. Three banks of light controls corresponded to the physical lamp placement throughout the performance space: front, center, and rear. Since her performances often took place in movie theaters and used the movie screen, this division could also correspond to top, middle, and bottom.

Significantly, Hallock-Greenewalt recognized that there is no inherent relationship between color and music, that color does not have an octave and projected light has no analog to musical harmony. This contrasts with contemporaries such as A. Wallace Rimington who developed more restricted forms of visual music. But the Sarabet did prevent multiple colors from being shown at the same intensity at the same time; each position in her scale — starlight, moonlight, twilight, auroral, diurnal, and superbright — allowed for only one lamp set at that level.

Nourathar performance was environmental, rather than image or symbol-based. While it was possible to pair these lamps with gobo templates to give shape to the light, Hallock-Greenewalt instead focused on fields of color, more like the abstract expressionists who were to follow, decades later.

Michael Betancourt is an artist, curator, editor, and theorist whose essays have appeared in *Leonardo*, *Semiotica*, *CTheory*, and other academic journals.

Make:
OPTICS
Makers know their eyes are as important as their hands. (That's why safety glasses are standard equipment.) In this special section, we bring new depth to the field of vision. We'll show you how to build your own kaleidoscope, 3D video camera, and webcam microscope, resurrect the opaque projector, and other light-distorting challenges. Look on!

HOMEBREW DIGITAL 3D MOVIES

Build your own stereo video camera and 3D viewer.

By Eric Kurland

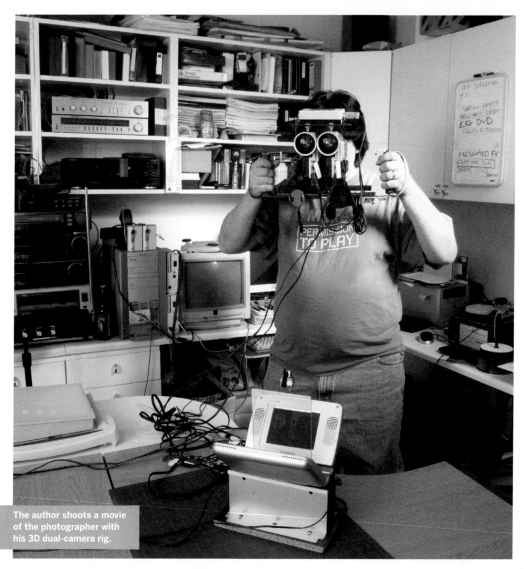

The author shoots a movie of the photographer with his 3D dual-camera rig.

Photograph by Amy Crilly

I have two eyes. And because of that simple fact, I also have *stereopsis*, the ability to perceive depth. When I was about 7 years old, I gazed into a View-Master toy and saw an amazing three-dimensional picture, and I was hooked. Today, I create 3D videos, using various homebrew camera rigs and displays. I'll introduce you to a few of my devices, but first, a quick history lesson.

In 1838, British scientist and inventor Sir Charles Wheatstone theorized that seeing with two eyes together is what allows us to see in 3D. Wheatstone deduced that each eye observes a slightly different view of the world, and our brain fuses these two perspectives together, interpreting the parallax differences as depth. He called his discovery stereoscopic vision (literally meaning "to see solid") and built an optical device, the stereoscope, that allowed three-dimensional viewing of pairs of drawings.

With the invention of photography, and later cinema, real-life images could be captured with two lenses and viewed in 3D. The popularity of stereoscopy has persisted over the years. In the 1890s, arcades offered 3D peep shows as entertainment, and the handheld stereoscope was a common item in home parlors — the TV of the Victorian era. The 1950s and 1980s both saw 3D movie "booms" come and go, due to the technical limitations of the times. And currently, in the age of digital video, stereoscopic 3D is seeing a major rebirth.

My own foray into 3D video began a few years ago, after I attended the monthly meeting of the Stereo Club of Southern California. Many of the photographers at the meeting had pairs of digital still cameras mounted side by side for shooting 3D photos, and it occurred to me that I could build a similar hand-held rig for use with small camcorders.

Shooting 3D

Starting with a pair of Sony Handycams, I set out to build a stereoscopic rig. My plan was to make the distance between the lenses, called the *interaxial*, equal to my *interocular*, or the distance between my eyes. This would give a natural-looking 3D depth to my footage, and would allow me to view 3D while shooting, just by looking through both camera's viewfinders. Putting the lenses so close required removing the hand strap from the left camera.

I attached the cameras to a metal bar using quick-release mounts for easy removal, in order to access the tape and battery compartments. I fashioned a bracket from some spare parts to hold both cameras securely at the top and keep the lenses aligned. Inspired by director Mike Figgis' steering wheel-like camera stabilizer (the "Fig Rig"), I bolted together a pair of photographic flash bars with handgrips, salvaged from a flea market dollar bin, and created a "handlebar" stabilizer. This allows me full mobility with the rig, and puts the center of rotation between the two cameras.

To control recording, I use a device called the 3D LANC Master. Developed by Dr. Damir Vrancic of Slovenia, the 3DLM connects the cameras via the LANC ports and provides simultaneous control of most camera functions. It also keeps the video recording in sync by continuously polling the timing frequency of one camera, and adjusting the frequency of the other up or down to prevent drift. This is very important when shooting 3D, as any time disparity between the camcorders will result in nonmatching left and right views. Schematics and software for the 3DLM are open source under a GPL and are available for free.

Viewing Live 3D

With my camera setup complete, my next task was to build a portable stereoscopic video monitor, so others could watch live 3D during shooting. In movie theaters, stereopsis is achieved by projecting left and right images through two oppositely oriented polarizing filters onto a reflective screen. By viewing through 3D glasses made from matching polarizers, each eye sees only the corresponding projection. I decided to use the same principle for my monitor.

I started with two small LCD monitors capable of showing NTSC video, the kind that are strapped to the back of car headrests. The video output from each camera is input to one of these monitors. Because LCDs have a polarizing layer, these displays appear black to one eye and visible to the other when viewed through polarized 3D glasses. I found that the monitors had a clear plastic protective sheet glued over each LCD. These had to be carefully peeled up and removed, as they were depolarizing the light from the screens.

On one display I needed to flip the picture horizontally like a mirror image, so I opened the case and wired pin 62 of the PVI-1004C LCD controller chip to ground. I attached the LCD displays to each other at a 90° angle, their screens facing inward, and mounted a piece of half-mirrored glass between

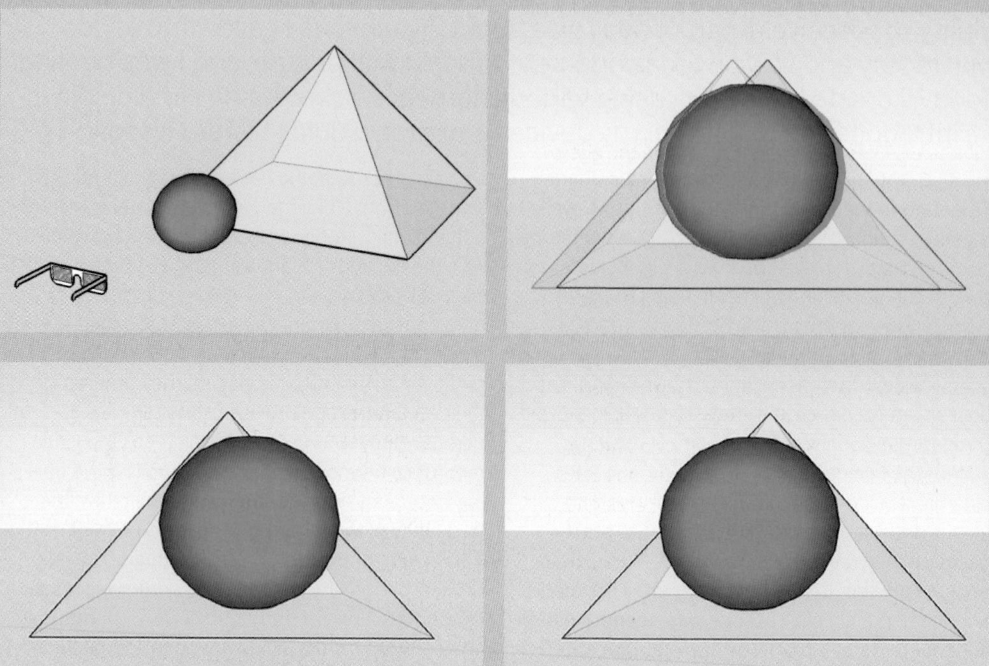

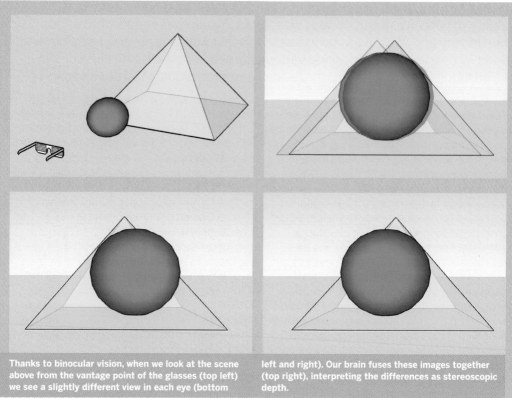

Thanks to binocular vision, when we look at the scene above from the vantage point of the glasses (top left) we see a slightly different view in each eye (bottom left and right). Our brain fuses these images together (top right), interpreting the differences as stereoscopic depth.

them. This glass superimposes the reflection of one screen on top of the other. When viewed through polarized glasses, the reflected image and its polarization are reversed, each eye sees only one screen, and we have live 3D video of whatever is being shot.

Editing 3D

Shooting with two cameras creates two individual video files, which are digitized into the computer for editing. First, I use the freeware application StereoMovie Maker, developed in Japan by Masuji Suto, to correct misalignments in my footage, which can cause eyestrain.

In StereoMovie Maker, I am able to load both the left and right videos and visually transform, scale, and rotate them while viewing in 3D with anaglyph glasses. Anaglyph is the method in which the two pictures are combined into a single image with one eye in red and the other in cyan. Primarily used in printed stereoscopy, anaglyph also provides a means of viewing depth on any computer screen using inexpensive red-cyan glasses.

Once satisfied with the alignment, I save my videos as a single file, formatted side-by-side in a split screen, and twice as wide as a normal video picture. I prefer this format, as it ensures that the two views always remain in sync throughout the editing process. The footage can be cut together in any standard video editing program. My system is PC-based, so I use Adobe Premiere, but the same techniques would apply to a Mac Final Cut Pro system. One thing to take into account when editing 3D is that drastic depth changes between consecutive shots can cause eyestrain.

To watch my completed movies in 3D, I use Peter Wimmer's excellent Stereoscopic Player program, a full-featured media player for stereo video files that converts on-the-fly to the many different viewing formats required by stereoscopic displays and projectors. Both Stereoscopic Player and StereoMovie Maker are Windows-only applications, but they will run on Intel Macs running Windows.

Showing 3D

In order to show 3D video to audiences, I have a dual-projector setup, just like the 3D theaters of the 1950s, using two projectors, polarizers, and a silver screen. The only real difference is that my projectors are small

Illustration by Eric Kurland

DLP digital models, and my "film" is a file played back by computer. This arrangement works well for large audiences, but I also wanted some method of carrying my 3D movies with me to show at a moment's notice — a portable stereoscopic media player.

The Sony PSP looked like it would be the answer. The PSP can play MPEG-4 files from a flash memory card and it has a nice, wide screen — wide enough to hold side-by-side-formatted left and right images.

In fact, the PSP is just about the same size as a standard vintage stereo card. I decided it would be cool, and somewhat steampunk, to mount a PSP onto a circa-1904 stereoscope.

As luck would have it, the PSP fit almost perfectly between the two card-holder wire clips. I didn't want to physically alter the viewer, as it's an antique, and I wanted the PSP to be removable, so I cut two loops of thin velcro strapping, just long enough to go around the PSP and hold it firmly to the slide bar.

To get my videos onto the PSP, I converted them to x264 compressed MPEG-4 files at the PSP's screen resolution of 480×272 pixels, and copied them to a PSP Memory Stick. Sure enough, side-by-side video files played on the PSP and viewed through the stereoscope's eyepiece are seen as a single three-dimensional movie.

The whole setup works perfectly. I can easily carry around a bunch of my homemade 3D movies on a Memory Stick in my pocket, and quickly show them to people through the "PSPscope" — a perfect marriage of 19th- and 21st-century technologies.

1. Build the 3D dual-camera rig.

1a. On one of the video cameras, remove the hand strap. This will be the left-eye camera.

1b. Attach the quick-release mounts to the bottom of the camcorders. Make sure that the mounts are perfectly straight, and tighten them down with a screwdriver or a coin. Position the cameras as close to each other as possible on the twin-camera bar, and tighten the thumbscrews until both cameras are rigidly secure (Figure A, next page).

1c. Optional: If you have the hot-shoe microphone extenders, place them on the cameras and tighten the thumbscrews. Attach the extenders to the 6" flat steel bracket by placing bolts through the bracket and tightening a wing nut onto each bolt (Figures B and C). The mic extenders should be parallel. Make sure that the camera lenses are aligned.

MATERIALS

FOR THE 3D DUAL-CAMERA RIG:

- **Matching pair of compact video cameras** Any pair of camcorders will work, but the 3D LANC Master will maintain sync only on certain older Sony MiniDV cameras. Mine are Sony DCR-PC100s.
- **Twin-camera mounting bar** from a specialty photography store
- **Quick-release camera mounts (2)**
- **3D LANC Master camcorder controller** Build your own from the schematics at dsc.ijs.si/3dlancmaster or order one pre-assembled from inddd.com.
- **Screwdriver**
- **Hot-shoe microphone extenders (2) (optional)** These were an accessory with the camcorder microphones.
- **6" flat metal bracket (optional)**
- **Adhesive velcro squares (optional)**
- **Camera flash bars with handgrips (2) (optional)** Any 2 should do; mine are a mismatched pair from a flea market.
- **Dog training clicker (optional)**

FOR THE LIVE 3D MONITOR:

- **Audiovox EX50 5" LCD video monitors (2)** You should be able to use other LCD panels, but you'll need to check the polarization.
- **5"×7" half-mirrored glass** aka flat glass beam splitter. You can use a cut piece of window glass but the results may not be as good.
- **Short length of sheathed wire** such as telephone wire
- **Self-adhesive velcro squares**
- **½" velcro strapping**
- **10"×6" piece of plywood**
- **1½"×8½"×3½" metal L-brackets (2)** I had these in a scrap parts bin; you'll need to find a similar substitute or perhaps use wood blocks instead.
- **½" wood screws (6)**
- **Linear polarized 3D glasses** There are 2 types of polarized 3D glasses, linear and circular. Only the linear kind will work with these LCD panels.
- **Precision screwdriver set** Phillips and flathead
- **Fine-tipped soldering iron and solder**
- **Magnifying glass**

FOR THE PSP STEREOSCOPE VIEWER:

- **Sony PSP**
- **Holmes-style stereoscope** vintage or new
- **½" velcro strapping**
- **Scissors**
- **Pliers**
- **Adhesive rubber foot or nut and bolt (optional)**
- **Electric drill (optional)**

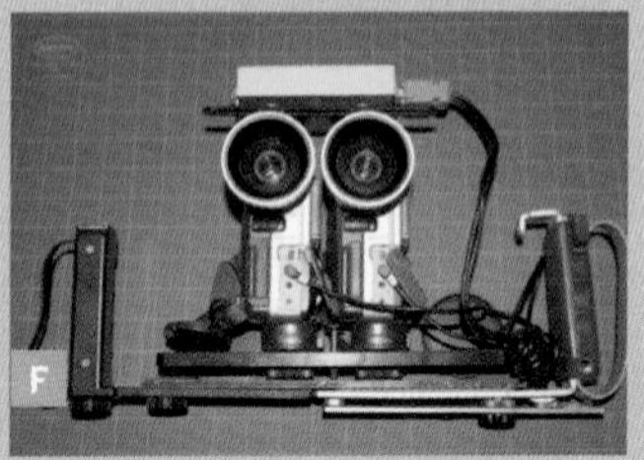

1d. Optional: Create a camera stabilizer by placing the flash bars end to end and attaching them to the bottom of the twin-camera bar with the handgrips pointed upward (Figure D).

1e. Remove the cameras from the bar. Insert blank tapes and charged batteries. Place the cameras back on the bar. Use velcro to attach the 3D LANC Master to the rig (Figure E).

1f. Connect the 3D LANC Master cable to the cameras' LANC ports (Figure F). Power up the Handycams, use the 3D LANC Master's controls to reset the cameras and get them in sync, and you're all set to record.

TIP: If you can't get a 3D LANC Master, you can still shoot 3D video with your twin camcorders. Use a dog training clicker to make a sync "pop" on your soundtrack. In editing, you can use the clicker peaks on the audio tracks to align the videos. Just be aware that on long takes your scenes may drift out of sync.

2. Make the live 3D video monitor.

2a. Open the case of one of the LCD monitors. First, use a small flat-head screwdriver to pop the rubber feet out of their holes, and remove the 4 Phillips-head screws from the case; 2 of these are under the hinge, so rotate the hinge until you see the rubber feet. Make sure to note which screws go where, as they are different sizes (Figure G).

2b. Carefully separate the 2 halves of the case. Remove the 2 Phillips screws holding down the printed circuit board (PCB) below the screen (Figure H). Very carefully flip over the PCB. Locate the LCD timing controller chip, a 64-pin IC marked "PVI-1004C." Using a magnifier, find pin 62 and follow its trace to a small board-mounted resistor. Carefully solder a short piece of wire from this resistor to the ground point on the PCB, next to the ribbon connector (Figure I).

2c. Close up the case, replacing the screws and rubber feet.

2d. Using the thinnest flat-head precision screwdriver,

Photography by Eric Kurland

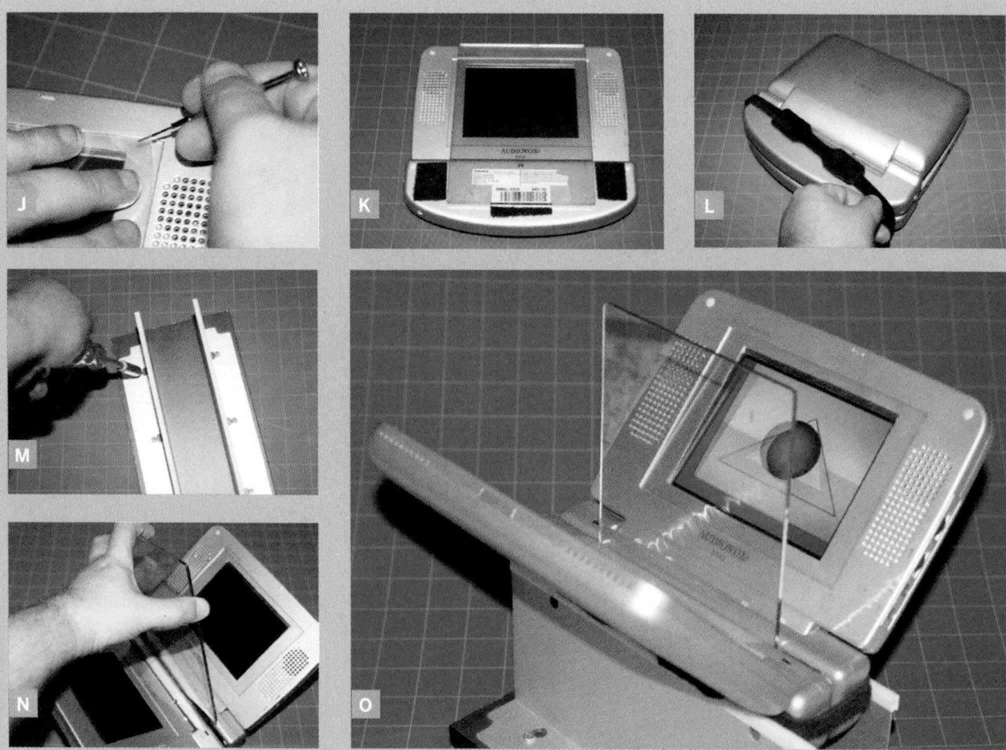

pry up a corner of the protective plastic sheet covering the LCD screen (Figure J). Slowly peel the plastic from the case. Repeat with the second LCD.

2e. Place the half-mirrored glass against the hinged base of the first monitor so that it extends over the display. Peel and stick several velcro squares onto the base around the glass (Figure K). Repeat with the second monitor, using the opposite halves of the velcro squares.

2f. Remove the glass and align the second monitor so that it faces the first. Press the 2 bases together, fastening the velcro.

2g. Cut a length of velcro strap long enough to encircle the base with a few inches of overlap. Wrap the strap around the base tightly and fasten it to itself (Figure L).

2h. Make a platform to hold the monitors. Using wood screws, fasten the L-brackets to the plywood so that they create a 2¼" opening (Figure M). Fit the monitors into the opening so that they stand upright, then bend each one back on its hinge 45°, so that the 2 form a 90° angle.

2i. Slide the glass between the bases of the LCDs (Figure N). It should clear the velcro squares and fit very snugly.

2j. Attach the power supplies, connect the video inputs of the LCDs to the video outputs of the cameras, and power everything up. Look down through the glass at one LCD, and you should see the other display reflected in the glass (Figure O). Tweak the angle of the monitors until the images are perfectly superimposed. Adjust the brightness levels of the panels until they match.

2k. Put on some 3D glasses and you can watch live stereoscopic video.

3. Make a PSP stereoscope.

3a. Make sure your PSP has the latest system firmware that supports playing full-screen MPEG-4 files. To check for this, go to the PSP's system menu and select Network Update.

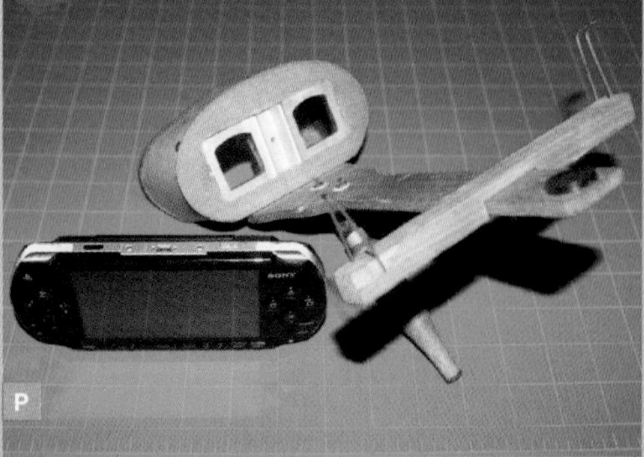
P

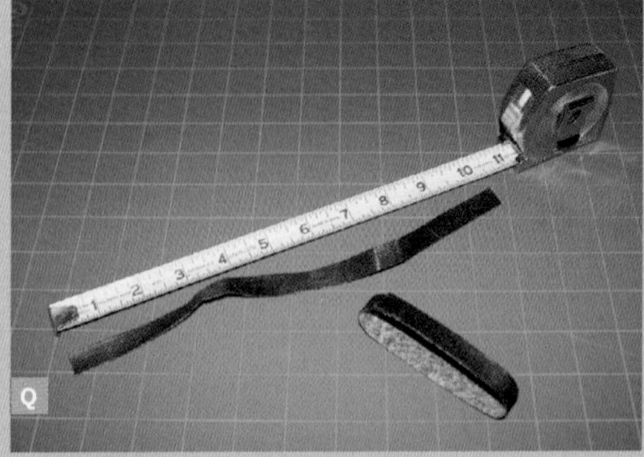
Q

R

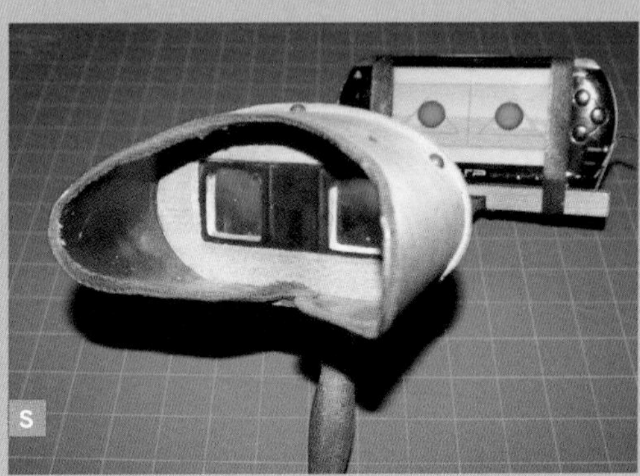
S

3b. Cut two 10½" lengths of velcro strap. Loop each piece, and overlap the ends 2½", fastening securely (Figure Q).

3c. Place the PSP on the sliding bar of the stereoscope, between the wire card-holder clips. If the PSP doesn't fit, use pliers to bend the wires.

3d. Holding the PSP firmly in place, slide a velcro strap over one end of both the PSP and the sliding bar (Figure R). The fit should be tight, but still allow you to move the strap on and off easily. If necessary, adjust the overlap to tighten or loosen the strap. Fit the second strap on the opposite end.

3e. Position the straps right at the edges of the screen without obscuring it (Figure S).

3f. Optional: Because of the weight of the PSP, the sliding bar has a tendency to slide off the back of the stereoscope. To prevent this, stick a rubber foot on the main bar, near the end.

3g. Save your 3D videos as PSP-compliant MPEG-4 files, using x264 compression, at a resolution of 480×272 pixels. Place your files into the root directory of a Memory Stick Duo. Fire up the PSP, load a side-by-side stereoscopic video, and enjoy a 3D movie.

RESOURCES

Stereo Club of Southern California: la3dclub.com

3D LANC Master free schematics and software: dsc.ijs.si/3dlancmaster

StereoMovie Maker free download: stereo.jpn.org/eng/stvmkr

Stereoscopic Player: 3dtv.at

Eric Kurland is an award-winning filmmaker and digital artist, currently developing several independent stereoscopic projects. In his spare time, he sometimes wears a gorilla suit and space helmet. Visit his 3D blog at retinalrivalry.com.

RECYCLED KALEIDOSCOPE

Make a classic optics toy from an old CD case.
By Carolyn Bennett

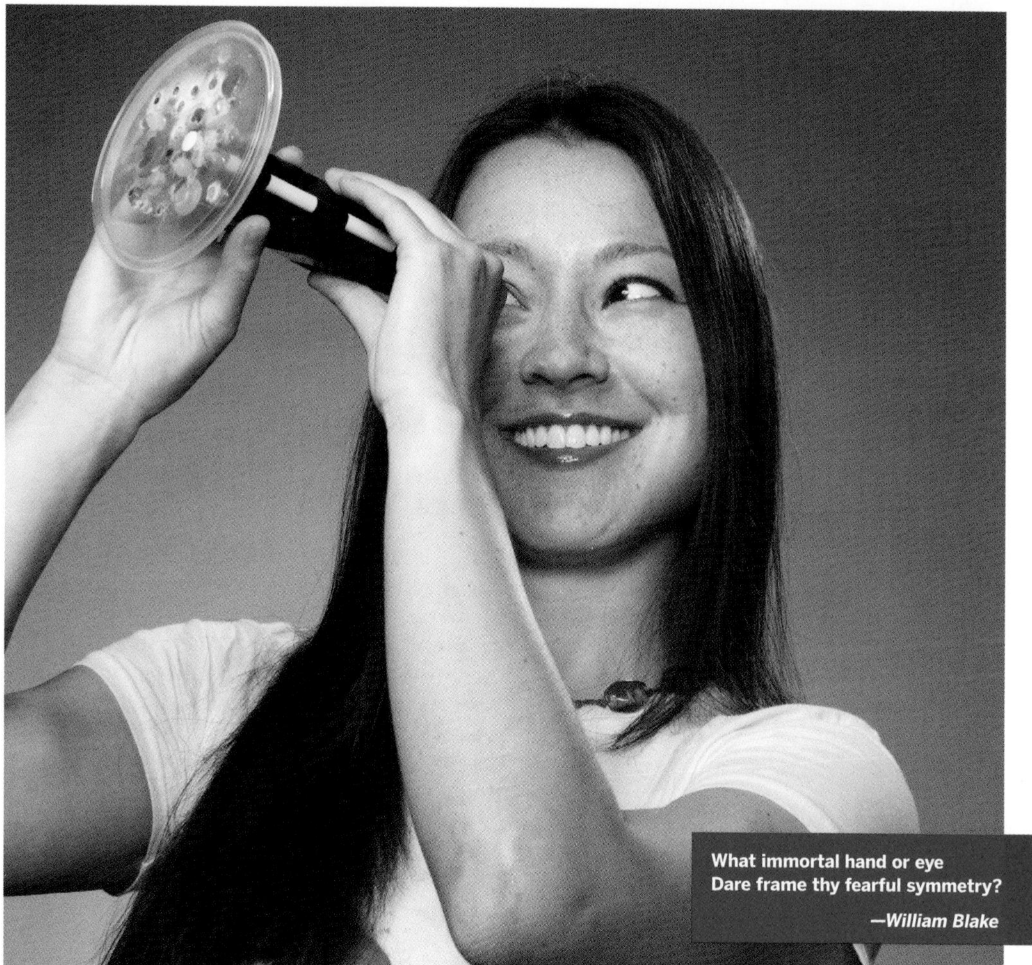

What immortal hand or eye
Dare frame thy fearful symmetry?
—William Blake

Photograph by Robyn Twomey

The kaleidoscope was invented in 1816 by a Scottish physicist named Sir David Brewster, and it has intrigued people of all ages ever since. Through the years, kaleidoscopes have been made of nearly every possible material. Now it's time to take the kaleidoscope green. Here's a simple one you can create from recycled materials and common household items. For the mirror elements, we'll use pieces of an old "jewel box" CD case backed with black paper or neoprene.

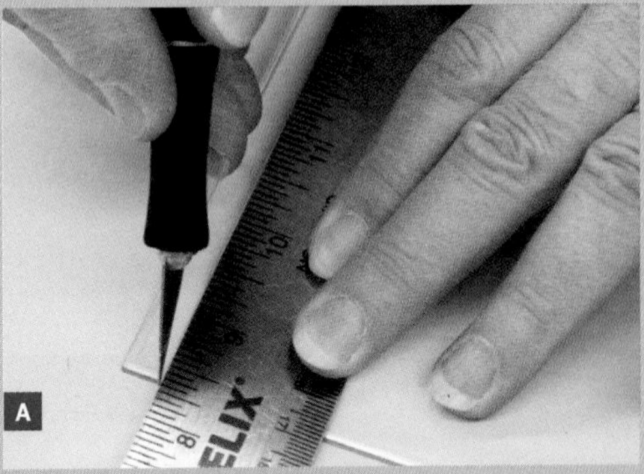

1. Remove all the raised edges from 1 half of the CD case, to get a flat piece of plastic (Figure A). It's easy to cut the plastic, if you're careful. Score along the inside edges of the CD case with an X-Acto knife, running it along each edge about 6–8 times. This will weaken the plastic. Then snap the ends off with your fingers.

MATERIALS

Clear CD case Try to find one that isn't scratched.
Pencil with eraser
Pushpin or thumbtack
Scotch tape
Strong tape such as electrical, cloth, or foam tape
Clear plastic lid like from a yogurt container
Craft glue
Beads, stickers, confetti, and assorted small treasures
Rubber bands (2)
Black paper or black neoprene/foam rubber
X-Acto knife
Ruler preferably metal
Small saw (optional)

2. Cut 3 strips of case plastic that are exactly the same width. My easy method was to use the width of my ruler as the mirror width, scoring along each edge and then measuring the next strip from the previous edge (Figure B). Score the plastic over and over; it will become so soft that if you keep scoring you could cut through it completely. Instead, align the score lines with the edge of a table, and snap the strips apart cleanly.

3. Form an equilateral triangle shape (technically a prism) with the 3 strips of plastic. Use the rubber bands to hold the prism tube together, then wrap around it tightly with tape to secure it (Figure C). You can remove the rubber bands after it's taped.

4. Cut a piece of black paper or foam that's the length of the triangular prism tube and wide enough to wrap around it completely, with a little overlap. Glue or tape this black backing around the tube to create the mirrors.

5. Measure and cut the pencil to the length of the prism tube plus enough to let the eraser hang over

Photography by Carolyn Bennett

the edge. You can cut through the pencil with an X-Acto, a utility knife, or a small saw. Make sure the cut end is smooth.

6. Use electrical, cloth, or foam tape to tape the pencil tightly to the middle of one of the flat sides of the tube, running parallel along the outside (Figure D).

7. Glue beads and treasures to the yogurt lid with craft glue. Allow the glue to dry overnight. It will dry clear, so don't be stingy; make sure everything is glued securely.

8. Stick the pushpin or thumbtack through the center of the decorated lid and anchor it into the middle of the pencil eraser.

That's it. Spin the disk and look through the tube to a beautiful image! Making this kaleidoscope unlocks the principles of how the instrument works.

Learn more about all things kaleidoscopic at brewstersociety.com.

Carolyn Bennett (cbennettscopes.com) has been creating kaleidoscopes for over 35 years. She also enjoys photographing antique objects and the farmlands that surround her home.

PRINT-AND-FOLD AMES ROOM

This classic illusion makes objects — and hobbits — seem to change size.
By Ranjit B.

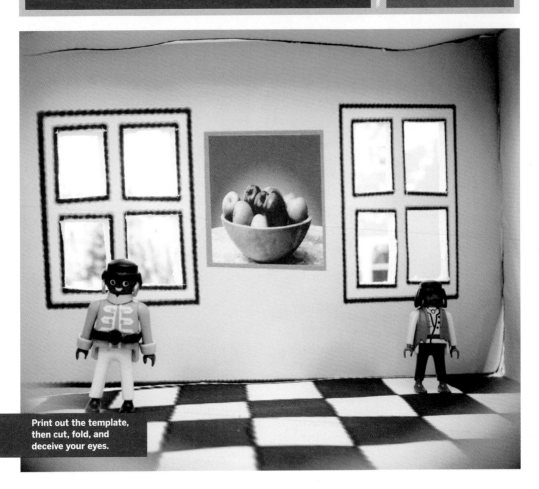

Print out the template, then cut, fold, and deceive your eyes.

An Ames room is a distorted box that creates an illusion, from one vantage point, of varying depth, distance, and size. Invented by American ophthalmologist Adelbert Ames Jr. in 1934, setups like this were used to make the hobbits look small next to Gandalf in the *Lord of the Rings* movies.

Here's how I constructed a miniature Ames room out of paper and cardboard. I'll start by explaining how I derived the template that I used, which you can download at makezine.com/14/amesroom. You can make your own template using this explanation or just use my template to make your own room.

Photograph by Sam Murphy

1. Derive the Ames room template (optional).

The goal is to make a room that looks perfectly even and rectangular from 1 viewpoint. You'll start by drawing a box and your viewpoint, and then ray-trace the corners of the box away to find new, distorted planes that will fit together along the same sightlines. I did this using Google's free design software, SketchUp (though you can use any 3D graphics program).

First draw a rectangle and use the Push/Pull tool to create a 3D rectangular room. On the menu View ⇒ Face Style, choose both X-ray and Shaded views. The room can be any size, but it's easier to design it smaller than you want your final room to be. I made mine 15" high by 25" wide by 30" deep.

Use the Freehand tool (Draw ⇒ Freehand or Command-F) to draw a viewpoint A, a few inches outside the center of 1 wall, the front wall; mine was 10" away. Switch to the Line tool (pencil icon, Draw ⇒ Line, or Command-L) and draw a window on the opposite, far wall.

Then draw a series of lines from A that intersect with and extend past all the corners of the room: B, C, D, E, F, G, H, and I. These are the sightlines you want to maintain (Figure A).

I decided to make the right side of the room the farther one, so objects would seem to shrink as they moved from left to right. To do this, mark a point D2 about 60" from A along the projected line AD, to the upper right corner. From there, drop a vertical line to AE and mark the intersection as E2, for the lower right corner. Then repeat the procedure on the left side, dropping a vertical line between B2 and C2 that connects AB and AC.

These 2 vertical lines determine the scale of the illusion, and D2-E2 should be twice as long as B2-C2. My D2-E2 on the right measured 20", so I found a point for B2 that made B2-C2 measure 10".

Draw horizontal parallel lines B2-B3, C2-C3, D2-D3, and E2-E3 to meet 4 new points on the sightlines that project through the near corners of the room, AH, AI, AF, and AG, respectively. Whatever distance these turn out to be is fine.

To complete the outlines of the Ames room, draw vertical lines B3-C3 and D3-E3. Both the ceiling and the floor are sloped; the rear and long-side walls are trapezoids, while the peephole and short-side walls remain rectangular (Figure B).

Now you need to obtain a completely flat view of each face, and export it as a separate 2D graphic. Begin by switching SketchUp to Parallel Projection.

Generating false perspective shapes with SketchUp. Fig. A: Apparent box and sightlines. Fig. B: Room wireframe. Fig. C: Flat view of one wall.

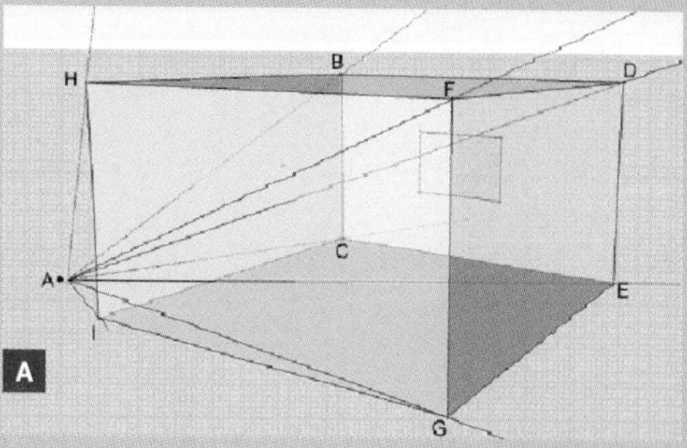

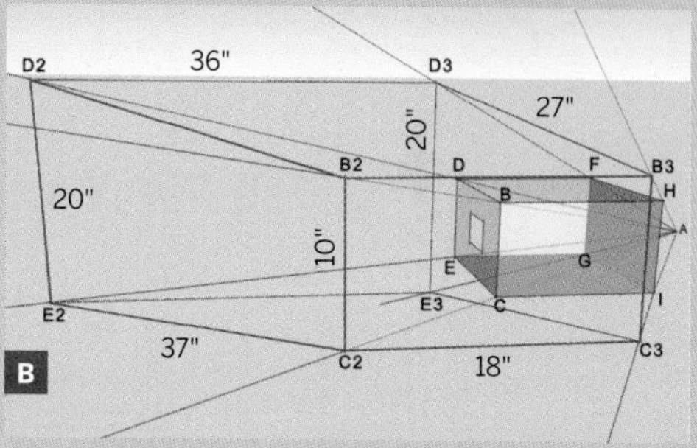

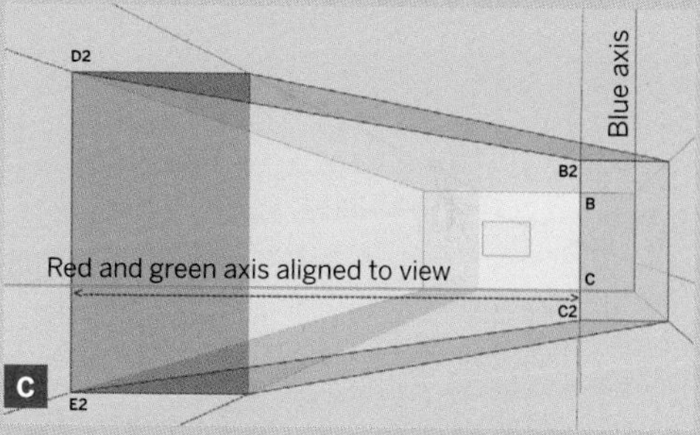

MATERIALS

Computer with Photoshop or other image editing software

Google SketchUp (optional) or other 3D graphics software, if you want to make your own template. Free download at sketchup.com.

Printer The larger the format, the better.

Cutting blade

Ruler

Stiff paper or cardboard

Adhesive tape or glue

Small props such as keys, pens, playing cards, toy figures

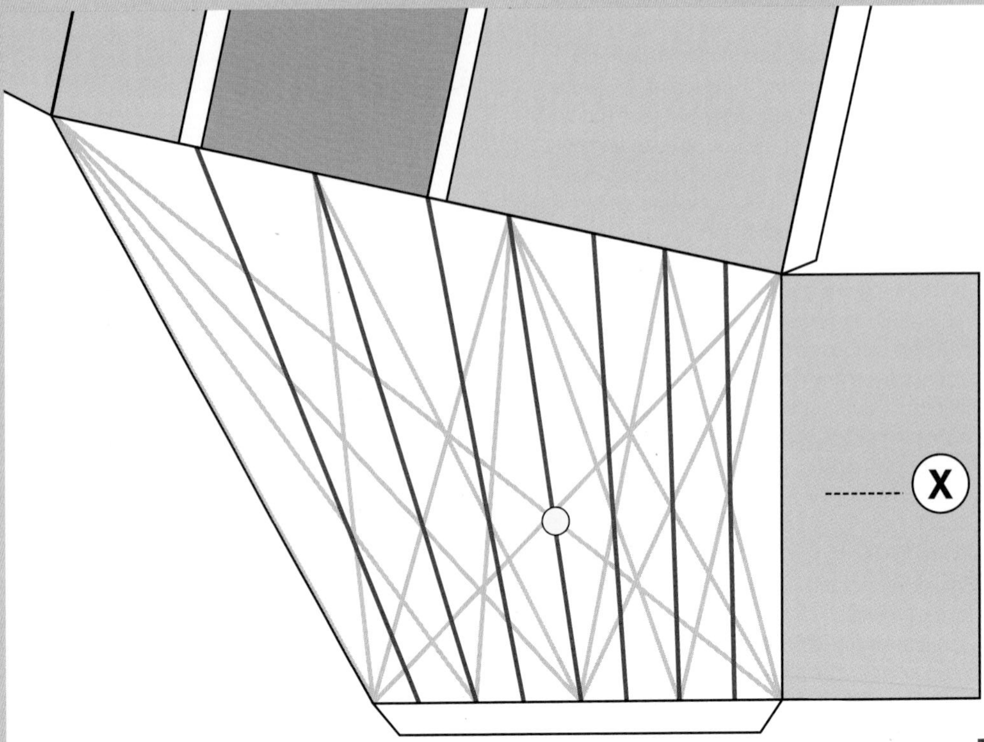

Use Camera ⇒ Standard to display flat, frontal views of the room's non-skewed faces (the side walls).

For the skewed ones, use the Rotate tool to show the face such that its width and height are at their maximum. In this position, 2 of the axes will be aligned, and 1 edge of the original cube will be aligned directly behind the corresponding edge of the face.

For example, a flat view of face D2-E2-C2-B2, the back of the room, results when the red and green axes are aligned, and line B-C is directly behind line B2-C2 (Figure C). My floor and ceiling shapes turned out to have corners measuring 45°, 105°, 75°, and 135°, while my trapezoidal walls had corners that measured 80° and 100°.

Export each flat face view as a 2D graphic file (File ⇒ Export ⇒ 2D Graphic). Then bring all 6 files into an image editor. I used Photoshop. Use some image editing skills to erase any text or unwanted lines, then bring all 6 clean faces together onto 1 large canvas. Orient and join the faces together to obtain your template. You might have to resize some of them so that everything fits.

After the shapes are joined, draw the windows, doors, and floor. Here's how to draw an 8×8 checkerboard tile pattern for the floor. First, find the midpoint on one of the short, undistorted sides. Then draw 2 diagonal lines across the floor from corner to corner. Draw a line bisecting the floor, from the side midpoint, through the point where the diagonals cross (the yellow dot), and across to the other side.

Follow the same procedure recursively to bisect each half of the floor, then all 4 quarters of the floor, for 7 total lines (Figure D). Then do it all over again from the perpendicular short side of the floor, to draw the other 7 lines of the grid.

Finally, add tabs to the outside edges of the template to help with taping the box together. Your template is complete.

2. Construct the room.

The larger the room, the more noticeable the illusion, so it's best to print it big. Use your template, or download mine at makezine.com/14/amesroom.

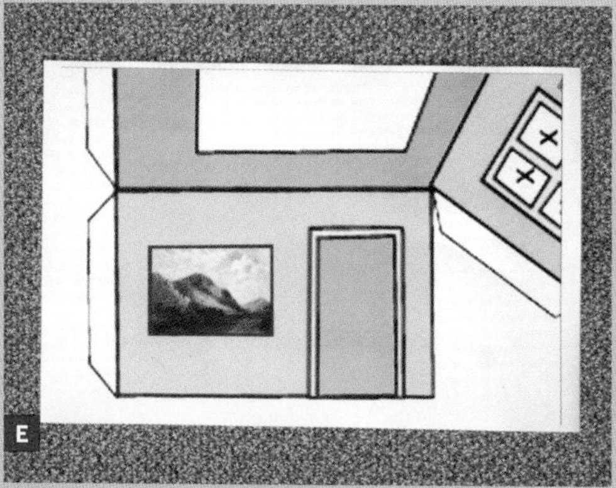

In Photoshop, I resized my final template to 48"× 54", but you can make it even larger, especially if your printer can handle poster size. The one I used could print up to 12"×18", so I used Photoshop to divide the image into 12 sections, and printed each one separately (Figures E, F).

I laid the sheets out on the floor, cut away the white borders, and taped them all together (Figure G). Then I cut out the windows, skylight, and peephole, marked with Xs on the template (Figure H). Finally, I folded in the tabs, carefully folded together the entire room, and secured all the edges with more tape (page 60, top).

3. Test the illusion.

The final step is to test the illusion using various objects such as pens, keys, playing cards, or small toy figures.

Place an object in one corner of the back wall and look through the peephole to view the inside of the room. Move the object slowly to the other corner and notice its apparent change in size. Identically sized objects, such as playing cards, that move across each other will seem to shrink and grow as they exchange places along the back wall.

Watch a video of Ranjit B.'s Ames room at makezine.com/14/amesroom.

Ranjit B. is a freelance graphic designer and web developer.

Microphotography on the Cheap:
For an event at the Children's Hospital where I work, I was asked to make an interactive display for kids. It had to be cheap enough to dispose of because there was no telling what kind of cooties it might transmit. I removed the front of an old webcam/kids' digital camera and glued it on top of a thrift store microscope. The results were fantastic using the original webcam software, the total cost was nil, and it even had the ability to take still shots if someone wanted a keepsake.

–John Dunlap

Photography by Ed Troxell

LENSLESS MICROSCOPE

A webcam's image chip is an ultrafine shadow-imaging stage.
By Tom Zimmerman

Imager chip from a 5MP camera mounted in a black box captures microscopic shadows of rotifer plankton.

Behind the lens of a webcam is an imager chip with thousands of tiny light sensors, each about 1⁄10 the diameter of a human hair. If you replace the lens with an LED light source, and place tiny objects on the imager chip, shadows will project onto the sensor, creating a lensless microscopic image. You can use the webcam's regular software to save pictures and video, or stream live images to the internet. Or use the imager chip from a security camera to see a colony of live plankton on TV.

Photographs by Sam Murphy (this page, A, B) and Tom Zimmerman (C, D, I)

A

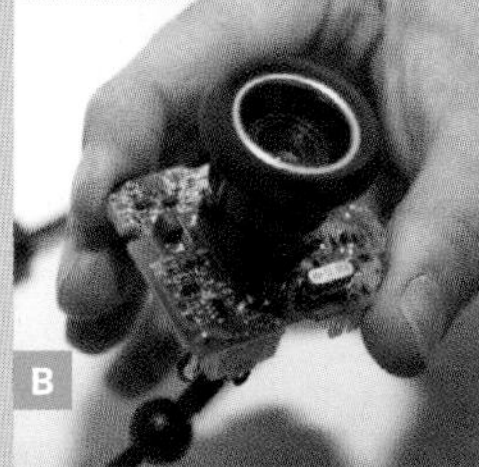

B

C

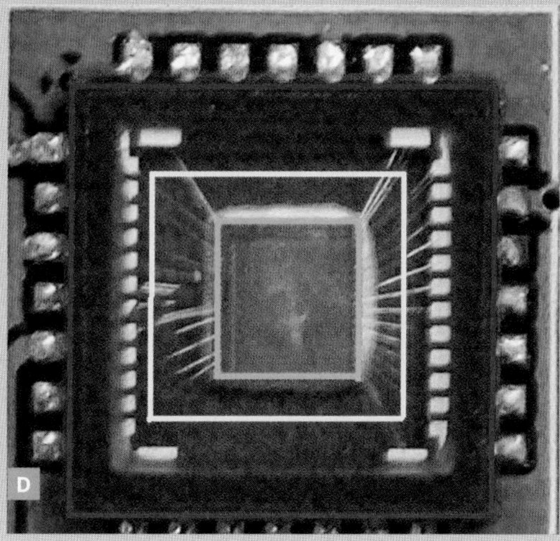

D

MATERIALS

Webcam for output to a computer. Higher resolution and a smaller image chip produce higher magnification. For output to a TV use Monochrome CCD video board camera, such as the PC302XS from supercircuits.com, $18.

2" black PVC end caps (2)

2"-diameter black PVC pipe, 8" long

Project box big enough to fit a camera board, RadioShack #270-1809

Silicone sealant for aquariums Regular silicone glue from a hardware store contains chemicals that kill organisms.

Resistor, 1KΩ RadioShack #271-1321

3mm blue LED Digi-Key #350-1562 from digikey.com Blue has the shortest wavelength, producing the sharpest image. Experiment with white and other colors.

SPST toggle switch RadioShack #275-612

9V battery

9V battery snap connector RadioShack #270-324

Corrugated cardboard, 3"×3"

Plastic, 3"×3"×1/8" thick if needed. See Step 5.

Blue masking tape

TOOLS

Scissors

Small screwdrivers slot and Phillips head

Hobby knife

Tweezers

Small square file

Hand drill with 1/8", 1/2", and 1" drill bits

1/4" drill bit approximately, depending on the size of your camera chip's imaging area

Soldering iron and solder

Hot glue gun

SPECIMEN

Live plankton available from Reed Mariculture, reed-mariculture.com

1. Remove the imager board from the webcam.

If you're using a video board camera, go to Step 2. Remove all screws on the web camera case, pry it apart, and remove the imager board (Figure A).

2. Remove the lens.

Unscrew the lens holder from the circuit board (Figure B) to reveal the imager chip (Figure C). Don't touch the protective glass cover; that's where your specimens are eventually going to go.

3. Prepare the dark box.

Examine the imager chip. The innermost square (Figure D, green outline) is the light-sensitive image die. The raised square just inside the soldered legs is the ceramic case (red outline). Measure out a third square area between these 2 squares, covering most of the fine gold wires that connect to the image die (yellow square). Pick a drill bit with a diameter close to this width. We'll be referring to this yellow square later.

Use the selected bit to drill a hole centered in the

bottom of the project box. Square the hole with the file (Figure E).

4. Prepare the imager.

You need to cover the chip's light-sensitive area with tape to protect it from silicone sealant. If the tape is too small, the image will be obscured by sealant. If the tape is too big, there won't be enough glass to glue to the dark box. Cut a piece of blue tape the size of the yellow square (Figure F) and center it over the imager chip (Figure G). As you cut and place the tape, hold it with tweezers so that you don't get finger oil on it.

5. Make a spacer.

If the imager chip is the tallest component on its circuit board, go to Step 6. But if other components stick up farther from the board's surface than the imager chip, you need a spacer.

In the center of the 3" x 3" piece of plastic, drill and file a square hole the size of the yellow square. Drill and file holes to fit over the contours of all tall components, so that the square hole will lie flat over the chip. Place a big wad of aquarium silicone sealant on the entire imager board; more is better (Figure H). Press the plastic spacer down onto the imager chip, centering the blue tape in the plastic's square hole (Figure I). Remove any extruded sealant for a better view. Alignment is critical, so do a careful job. Let the sealant set overnight.

6. Glue the imager into the dark box.

If you made a spacer (Step 5), squeeze a big wad of silicone sealant all over the spacer. Otherwise, squeeze it all over the board and imager chip (Figure H). More is better.

Place the circuit board in the bottom of the box, pressing it against the square hole and causing sealant to extrude out. Center the blue tape within the square hole (Figure J), removing any extruded sealant for a better view. Alignment is critical; the entire blue square must be aligned and centered within the box's square hole. The sealant will create a gasket to prevent water from reaching the circuit board. Let the sealant set overnight.

Find the imager's glass cover with the point of the hobby knife blade, and drag the blade around

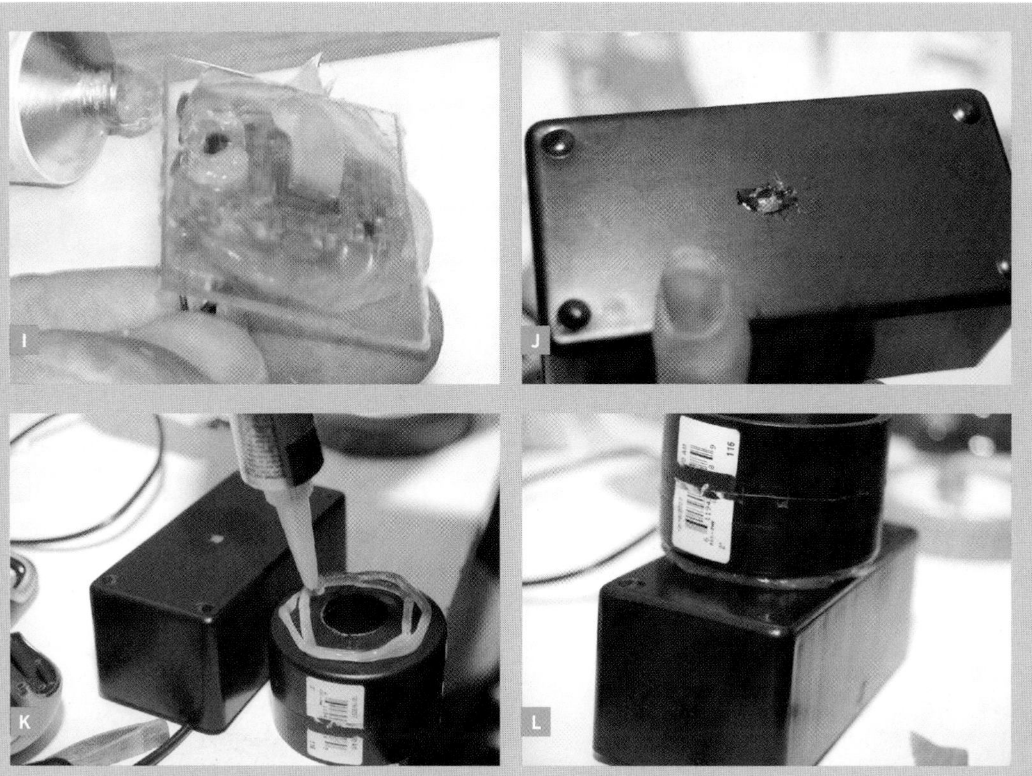

the perimeter of the square hole, cutting through the sealant and blue tape. Use tweezers to pull off the sealant and blue tape, revealing a pristine glass surface.

7. Prepare the end caps.

Drill a 1" hole in the middle of one of the PVC caps, then center it over the square hole in the dark box. Glue it down using silicone sealant (Figures K, L, M).

Drill a ⅛" hole in the center of the other PVC cap. Solder the LED, battery snap, resistor, and switch in series. If the LED doesn't light, reverse the battery leads. Press the LED into the ⅛" hole and hot-glue it and the other components to the top of the cap (Figure N).

Cut the cardboard into a 2" circle, drill a ½" hole in the center, and cut 10 or so radial lines ¼" inward from the circumference, spaced out approximately evenly, to make tabs that will fold back as you push the disk into the tube. Shove the disk 1" deep and flat into the bottom end of the PVC tube (Figure O). The inside of the tube is shiny and could cause reflections that would blur the image, but the hole in the cardboard only lets light coming directly from the LED reach the objects on the imager. This ensures that the shadows cast on the image sensing area are sharp.

8. Behold!

Put some salt or other small objects on the imager chip's glass cover, assemble the microscope with the LED end cap on top (Figure P), turn on the LED, and behold the microscopic world. Since we didn't touch the web camera's electronics, all of its software will work.

You can put solids (sand, salt, sugar), liquids (plankton samples, murky outdoor water), and objects (moth, fly) directly on the imager's glass cover. As with a regular optical microscope, the image comes from light shining through the subject. The magnification of the lensless microscope is the ratio of the monitor width to the imager sensor width (about 7mm).

Use a dry cotton swab to clean liquids off the glass. Use a straw to blow solids off (close your eyes!), or clean the glass with an alcohol-soaked swab.

Note that you'll get false colors due to imperfections in the imager's color filters and camera

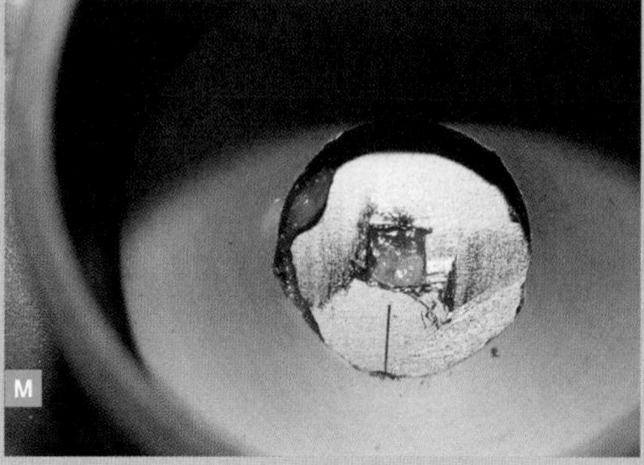
M

N

O

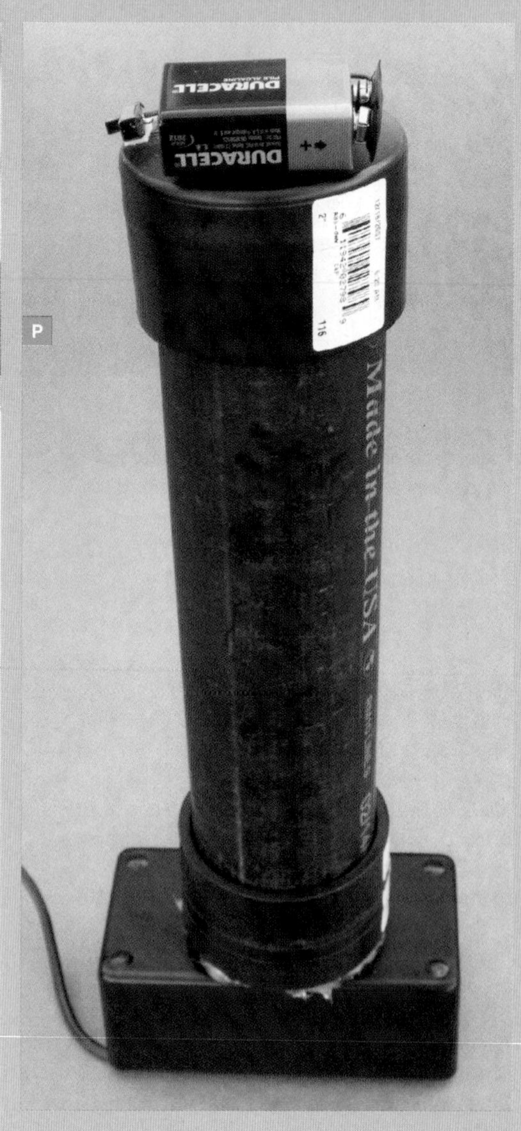

P

software. Try different-colored LEDs for interesting effects. If you're ambitious, you can hack a megapixel camera (Figure 5) in the same way and obtain higher resolutions — potentially much higher, depending on the camera.

See a video of rotifer plankton taken by the lensless microscope at makezine.com/go/plankton.

Tom Zimmerman is an inventor, educator, and researcher at the IBM Almaden Research Center who loves gadgets, LEDs, synthesizers, and hooking people up to computers.

Paint Brush Can Cover

An old CD makes a great temporary cover for an open can of paint, quart-size or smaller. Just cut it so it can slip over your paintbrush, and it will keep the brush suspended in the paint or solvent.

—*Frank Joy*

7 MILLIMETER WINDOW

Photos captured by the lensless microscope in various configurations.

1. Moth, VGA webcam imager, red LED.
2. Salt crystals, VGA webcam imager, white LED.
3. Copepod plankton, VGA webcam imager, blue LED.
4. Rotifer plankton, CCD black-and-white video imager, blue LED.
5. Portion of image of rotifer plankton, 5 megapixel imager, blue LED.

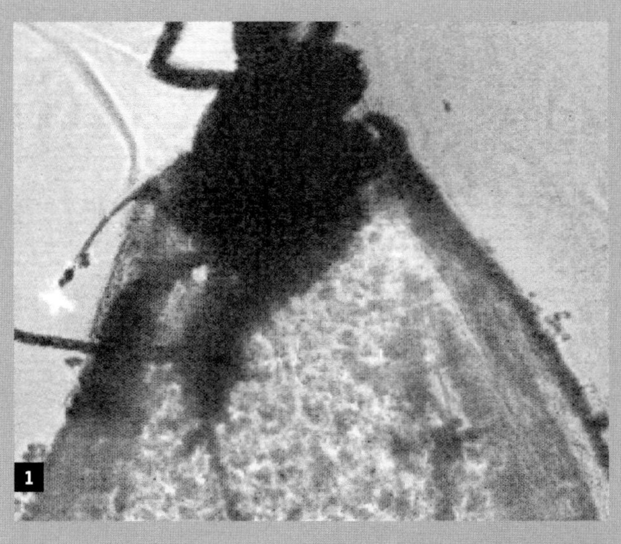

1

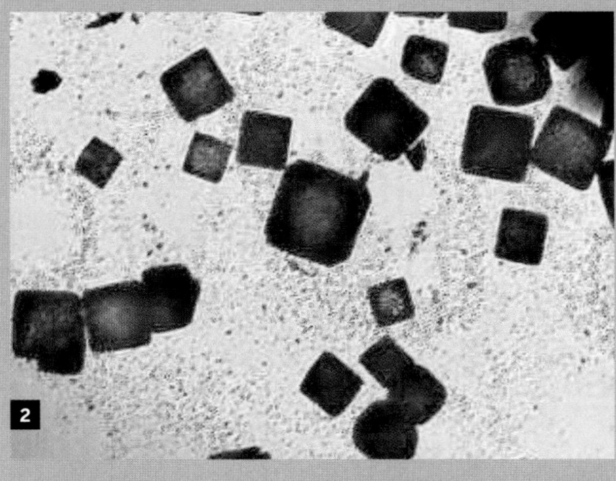

2

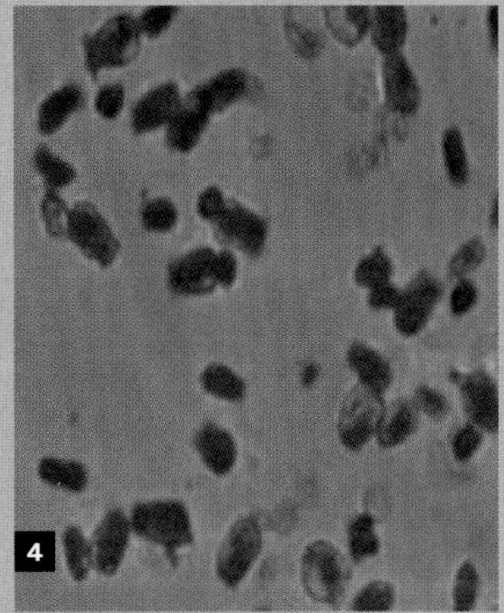

4

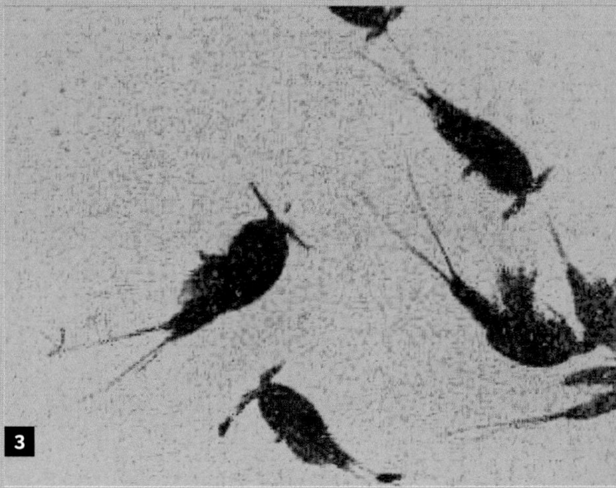

3

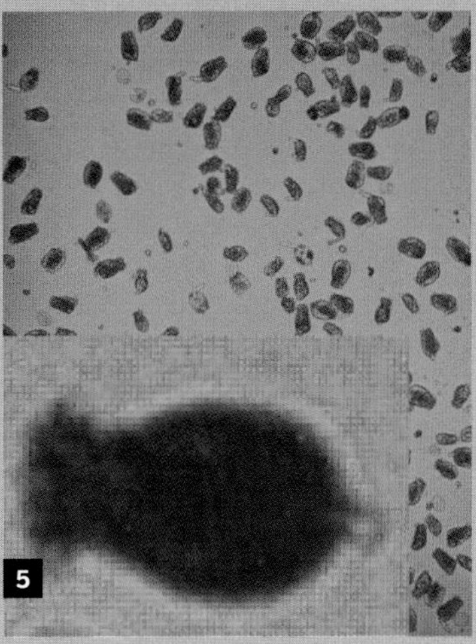

5

COSMIC NIGHT LIGHT

Make a glittering LED constellation jammed in resin — with no soldering!

By Kris DeGraeve

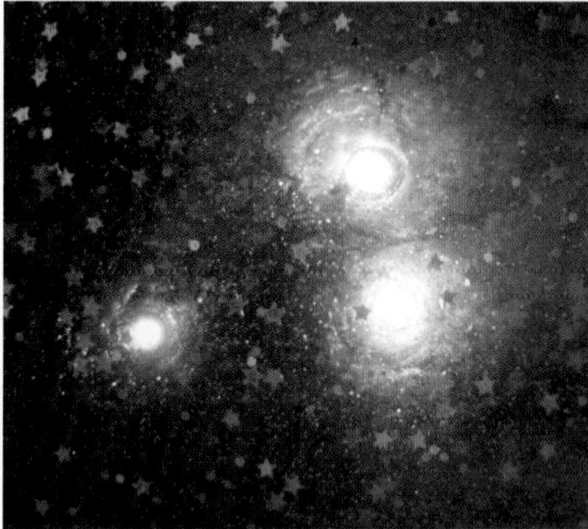

Photography by Kris DeGraeve

I wanted to make a night light with LEDs encased in resin that required no soldering — I can solder but I don't really like to. The project turned out to be one of my favorites, and beyond being a little tweaky getting all of the LEDs set, it's simple. The power comes from 2 coin batteries, so there's no risk of shock. And the finished product is a glossy, atmospheric light with a soft glow that looks great between my Martian lunch box and little plastic dudes landing on the moon.

1. Make a template.

First you'll need to choose a pattern for the LEDs, like a constellation. Most people would choose the Big or Little Dipper, so I went with Leo — it's the best, and it nicely fit the dimensions of the mold I had chosen. You can search online for a star chart, use my pattern, or just make one up.

To turn your pattern into a template, import it into a graphics program (I used Illustrator) and resize it to fit your mold, leaving a ¼" margin around the outside. Draw a small circle where each LED will be placed. Then draw 2 connect-the-dots lines that touch each LED on opposite sides, one for positive and the other for negative, making sure the lines never cross.

In an empty space, draw a 1" circle for the battery pack, another circle to fit the switch (mine was ¾") and a small rectangle for the resistor. Draw more lines and adjust positions to complete the circuit: the positive side connects to the positive battery terminal through the resistor and switch, and the negative side goes directly to the negative battery terminal. Place the switch where it will be easy to reach, and orient the battery holder to make it easy to change batteries. Make all lines thick and dark so they clearly show through your first layers of resin.

Finally, draw your mold's outline (3"×6" rectangle) around your pattern, flip the image (since you'll be using your template from the back), and print it out. I made 2 copies, with an extra one to refer to and make notes on. You can download my template at makezine.com/14/cosmic.

WARNING: Resin and epoxy are toxic, dangerous, and can damage many surfaces. Use only in a well-ventilated area, on a well-protected surface, and away from kids and pets. Read and heed all product safety warnings.

2. Cast the face layers.

You'll start at the face of the light and the bottom of the mold (Figure A, following page). Follow the instructions to mix up about ½oz of clear resin, and add glitter or other cosmic debris if you like. Catalyze it, pour into the mold, and tip the mold so resin covers the bottom and gets up onto the sides a bit. Let it harden completely.

I added a second layer of blue/green/yellow/pearly colored haze — don't do this if you want the LEDs to appear as sharp points of light. For the haze layer, mix another ⅓oz–½oz of clear resin, add glitter, and pour to cover the first layer. Add a drop of dye and swirl it with a toothpick. Add drops of different-colored dyes and swirl just until you like how it looks, then leave it alone until it's hardened. Mixing too much loses the effect. If you mess up, pour it out and try again. The reason you mix colors in this second layer instead of the first is that the toothpick can leave permanent marks on the mold.

MATERIALS

I rounded up most of these for cheap on eBay, but you can probably find them all at a craft store, a home improvement store, and RadioShack.

- Mold for resin **Mine was a 3"×6" rectangular mold from Castin'Craft, eti-usa.com.**
- 6oz–8oz of clear resin, and catalyst if needed **I used a polyester epoxy, but a 2-part one would probably be safer.**
- Resin dyes: black, pearl or white, and additional colors **I used blue and yellow to achieve colorful, hazy light, plus pearl or white, which helps the colors "pop."**
- Glitter **I used silver holographic and star-shaped glitter.**
- 5-minute plastic-approved clear epoxy adhesive
- Cyanoacrylate glue **aka super or Krazy glue**
- LEDs **I used 10, all 5mm white (besthongkong.com part #BEC0501044), but almost any should work.**
- Lightweight wire **Insulation isn't necessary; I used 24-gauge silver-plated beading wire.**
- CR2032 batteries (2) and coin cell battery holder mouser.com **part #534-1026**
- 150Ω resistor, ¼W or other value **from the calculator at led.linear1.org/led.wiz. This light was built using a 56Ω , ½W resistor for a 6V source, 3.3V diode, and 20mA current, but LEDs are forgiving, so you don't have to be too exact, and with a small array you can use just 1 resistor.**
- On/off toggle switch

TOOLS

- Alligator clip leads (2)
- Waxed paper
- Cardboard
- Small, strong spring clamps or binder clips (5)
- Masking tape
- Toothpicks
- Pliers **2 pairs is nice, but not necessary**
- Screwdriver

3. Test and attach the LEDs.

Load the batteries into the holder, attach alligator leads, and use them to test the polarity of each LED; it will light when you connect the positive voltage to the LED's positive side and negative voltage to the negative side. Mark the polarity for each. An LED's negative side is usually marked, but you should test them all anyway because you'd be bummed to make this project and then have the lights not work.

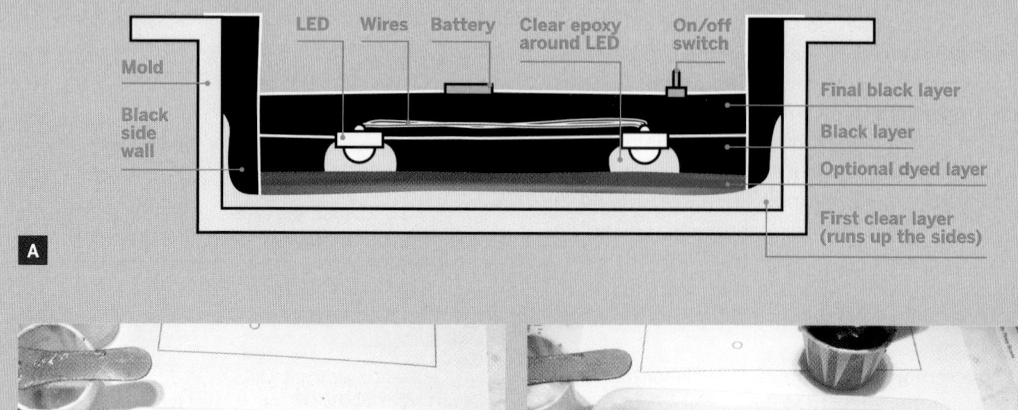

A

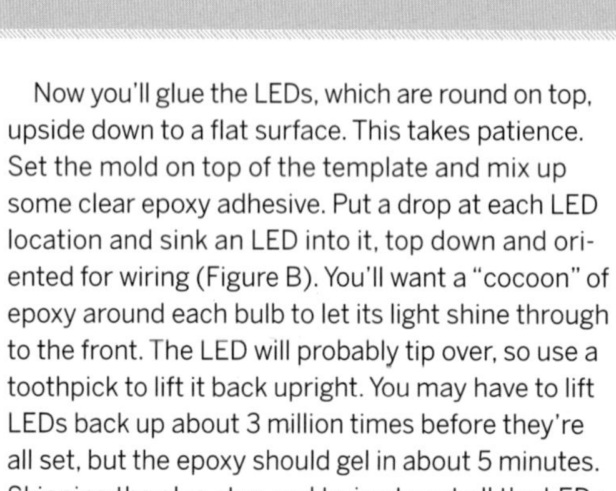

B

C

Now you'll glue the LEDs, which are round on top, upside down to a flat surface. This takes patience. Set the mold on top of the template and mix up some clear epoxy adhesive. Put a drop at each LED location and sink an LED into it, top down and oriented for wiring (Figure B). You'll want a "cocoon" of epoxy around each bulb to let its light shine through to the front. The LED will probably tip over, so use a toothpick to lift it back upright. You may have to lift LEDs back up about 3 million times before they're all set, but the epoxy should gel in about 5 minutes. Skipping the glue step and trying to set all the LEDs in a new layer of resin would give you very little time to fix the fallen ones.

4. Seal the LEDs.

Mix another ½oz of resin and dye it very black — the more opaque, the better. Pour small amounts in between the LEDs (Figure C) and let it spread. Then fill the mold up to a level that's just even with but not above the plastic backs of the LEDs (Figure D). This layer hides the wires, switches, and batteries from the front. Again, let this layer harden like you mean it — at least overnight, and until it clicks when tapped.

5. Build up the sides.

Now we'll build up the walls of the night light by pouring resin into the mold sideways. I've been working on this method for a while, and it lets you make a wide range of resin enclosures without having to use a 2-part mold.

Cut 2 strips of cardboard 1½" by 6½"–7". Stack and tape them together with masking tape, then wrap and tape waxed paper around the cardboard so that 1 side is smooth waxed paper with no tape.

Lay the mold on 1 long side with its rim hanging over the edge of a box or table. Use spring clamps to clip the cardboard strip to the rim, smooth side in, to enclose the low side of the mold. I used 3 clips along the bottom and 1 on each side (Figure E). The waxed paper should form a seal against the lip of the mold, and resin is viscous, so trust me, it'll hold.

Set the mold up over newspaper and run tape over the top to keep it from falling over. Mix up ½oz of black resin with optional glitter and pour it over the cardboard into the side of the mold, about ¼" thick. Let it harden, then unclip and peel back the strip, and repeat for the other 3 sides. I used ½oz of resin for each long side and ¼oz for each short side.

Illustration by Tim Lillis

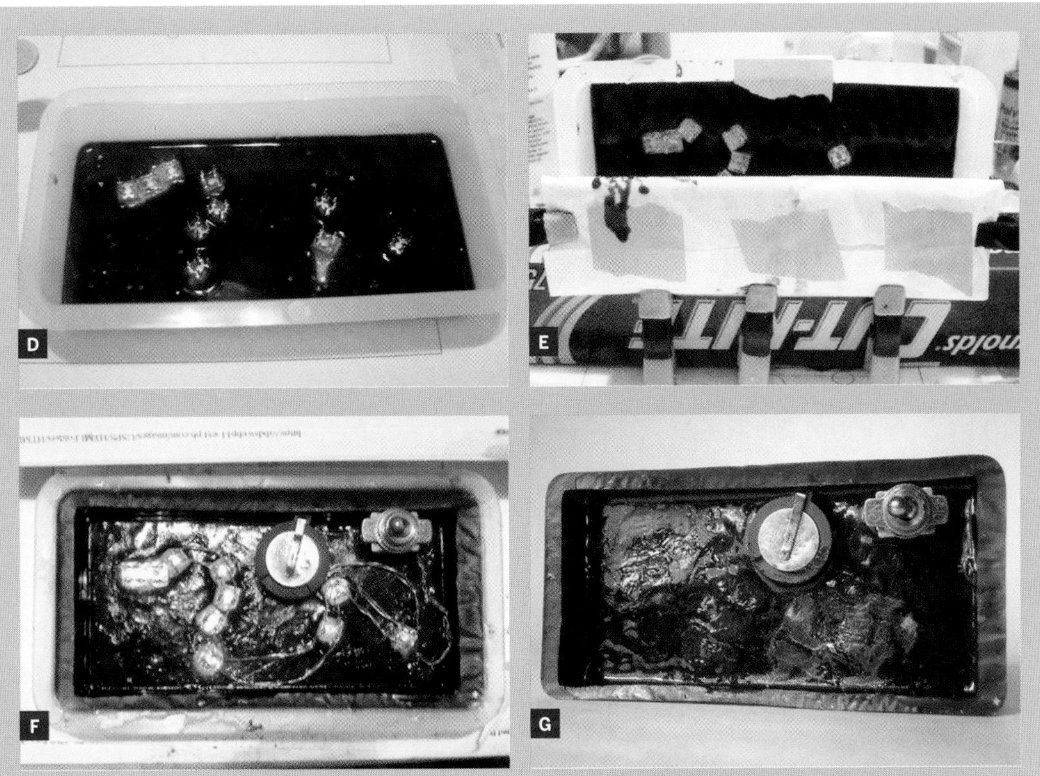

6. Wire it up.

Cut a 3' piece of 24-gauge wire, fold it in half, and hook it over the positive prong of the LED farthest from the battery (check your template). Wind the wire between the LEDs in overlapping figure-eights until you've connected all the positive prongs, then bend the prongs over the wire with pliers, squeezing tight. Repeat this process with the negative prongs. If an LED pops out, push it back in, finish wiring, and reglue it. Make sure the clear epoxy seals all the way around, or else black resin can seep in later and obscure the light.

At the end of the positive line, cut the wires off 1"–2" beyond the last LED and twist the ends with 1 end of the resistor. Place the switch and battery holder where you want them, and twist-connect the rest of the circuit with more wire: resistor to switch, switch to battery (+), and battery (–) to negative line. Trim any excess wire. You can super-glue loose connections to hold them until the final cast.

Carefully flip the switch on, and make sure all the LEDs light up (Figure F). Fix any that don't. Mix more epoxy adhesive and put a dollop over every LED. Keep the LEDs on and watch them while the epoxy sets; sometimes epoxy will slide between wires and take out a connection, but then you can turn the light back on by pushing with pliers or a toothpick. Again, make sure it's all really, really dry.

7. Pour the last layer of resin.

Mix up ½oz or more of dyed black resin and pour it into the mold to cover the LEDs and fill in under the switch and battery holder. Make sure you don't pour in so much that you disable the switch or make it impossible to remove the batteries. Let it harden.

8. Unmold, turn on, and admire.

Carefully pull away the sides of the mold and flex it until your light pops out. Handle it gently; even after the resin seems hard, it can take a few days to set completely. You might lightly sand the resin walls in back, but otherwise your light is done (Figure G). You can also cover the back with felt, secured by glue or velcro. Enjoy your new cosmic night light!

Kris DeGraeve is a full-time artist, designer, and maker. She posts about her current projects on technoplastique.squarespace.com.

WALL EYE

Build your own opaque projector.

By Steve Lodefink

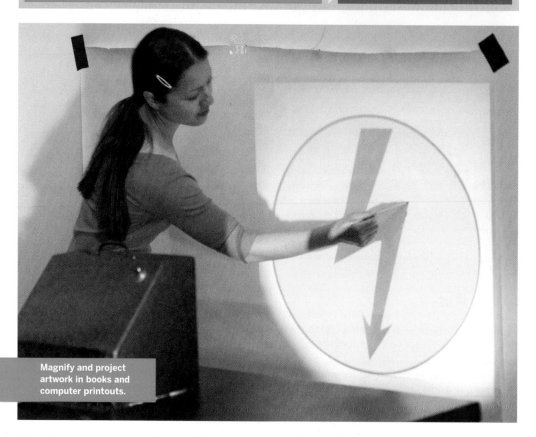

Magnify and project artwork in books and computer printouts.

Photograph by Robyn Twomey

What's an opaque projector? You know, one of those contraptions that takes flat, reflective subjects such as printed pages, leaves, or coins, and projects them onto a screen or wall. Opaque projectors were common classroom presentation tools during most of the 20th century.

Although made largely obsolete by the use of video cameras coupled to video displays, opaque projectors are still being made and sold. Marketed today mainly to art students and hobbyists for use as drawing enlargers, entry-level models tend to be dim little plastic toys that can only accommodate puny 3" or 4" originals, and have to be used in a totally darkened room due to their small apertures and weak light sources. If you're lucky enough to find one of the majestic 1,000-watt giants from yesteryear at a swap meet, then consider yourself charmed. I was not so lucky.

While searching for a projector to buy, I did happen upon a free PDF download of an old Edmund Scientific DIY booklet called "How to Build Opaque Projectors." Just 20 pages long, the pamphlet is packed full of projector theory, useful tables and formulas, design considerations, lighting and lens

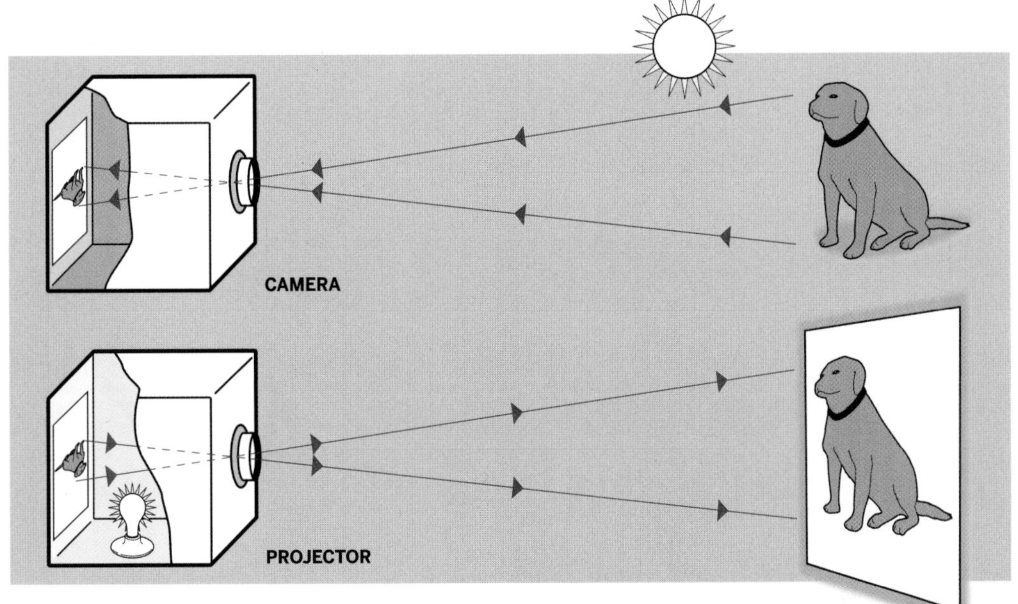

While a camera is essentially a box with the subject and light source outside and the reproduced image inside (top image), a projector is the exact opposite (bottom image), with the subject and light inside and the image projected outside.

systems, and projector variants. Not only was I impressed with the substance of this booklet, which left me feeling fully qualified to build a projector, but I was really taken with the style as well. The old-school draftsmanship and hand lettering were straight out of the postwar heyday of home engineering.

I thought it would be neat to build a piece of obsolete technology straight from a castoff, old-school set of instructions, so I decided to build my own drawing projector.

Making the Projector

A projector like this is really just a camera in reverse: a box with the subject and the light source on the inside, and the reproduced image on the outside.

Lens Selection

To make a projector, you start with a lens. The Edmund booklet details the pros and cons of various common lens designs of the day, balancing performance against cost and availability. There are 4 main characteristics to consider in a projection lens.

Focal Length The effective focal length for our purposes is the distance from the front element of the lens to the copy board while it's in focus. You want a lens with an FL somewhere between 6" and 12". An easy way to estimate the FL is to stand below a ceiling light fixture with the lens in one hand and a sheet of white paper in the other. Move the lens up and down until the image of the light fixture appears in focus on the paper. The distance from the paper to the lens is the FL.

Speed The f-ratio, or focal ratio, of a lens is the FL divided by the diameter of the aperture. Lenses with lower f-numbers are said to be "faster" because they are open wide and let in more light, while "slow" lenses have higher f-numbers (smaller maximum apertures). You want the fastest one, i.e. the lowest f-number.

Field of View A wider field of view will cover more copy area at a given FL, allowing you to project larger originals.

Field Distortion Find a lens that produces as little curvature of field as possible. Also look for one that exhibits good focus all the way out to the edges of the field.

Fortunately for us today, there is a glut of take-out copy machine lenses on the surplus market. These tend to be fast Cooke triplet lenses with nice wide field of view, great focus across the field, and little appreciable field curvature. They're tailor-made for this project! I got a 10" copy lens from American Science & Surplus (sciplus.com) for about $15 including shipping. Another good option might be lenses from old projection TVs.

Illustration by Tim Lillis

A

The Cabinet

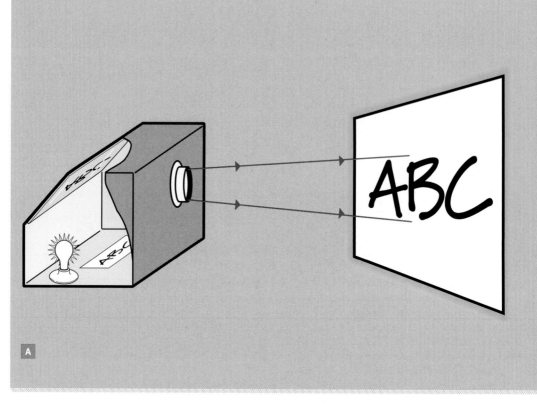

The finished projector is basically a plywood cabinet with a lens holder in the face, an internal lighting system, and an angled front-surface mirror behind the lens (Figure A).

The mirror serves to flip and invert the image so that it will appear right-reading on the screen, and to "fold" the optical path, allowing the cabinet design to be more compact, and the copy to be laid flat. The bottom of the box is left open and the projector simply placed on top of the copy. Focus is achieved by sliding the lens in and out in its carrier.

The dimensions of your projector cabinet will depend upon your lens selection, expected copy size, and desired magnification. A quick way to come up with a rough layout is to cut out a paper pattern that represents your "light cone." The shape will be an isosceles triangle with the base being the maximum width of your copy area, and the height being the effective FL of your lens. Fold the paper triangle at a 90° angle roughly at the midpoint. The fold will describe the correct position of the mirror.

Now build your cabinet to dimensions that can accommodate this pattern. I used scrap and salvaged ⅜" plywood for the back, top, and sides, and some ½" ply for the front panel.

Assembly

After gluing up the box, I made a lens carrier from an empty cocktail peanut can with the bottom cut off. The size and price were about right, and the metal rim provided a nice finished look when the can was epoxied into a circular hole cut into the front panel. To get a nice snug fit in the carrier, I tightly wrapped the lens barrel in several turns of black yarn.

For lighting, I installed 2 ceramic lamp bases high up in the cabinet, 1 on each side of the lens, and made reflectors out of some mylar insulation material. I wired them together with a lamp cord, which exits the back of the cabinet. I'm using a pair of 23W compact fluorescent bulbs, which give off the equivalent of 200W of tungsten light; this is still pretty dim, but I wanted to avoid the heat produced by brighter incandescent lamps.

A strategically shaped bank of high-output LEDs would nicely augment these lights. The entire inside is painted black to prevent glare from stray

Photography by Steve Lodefink; illustration by Tim Lillis

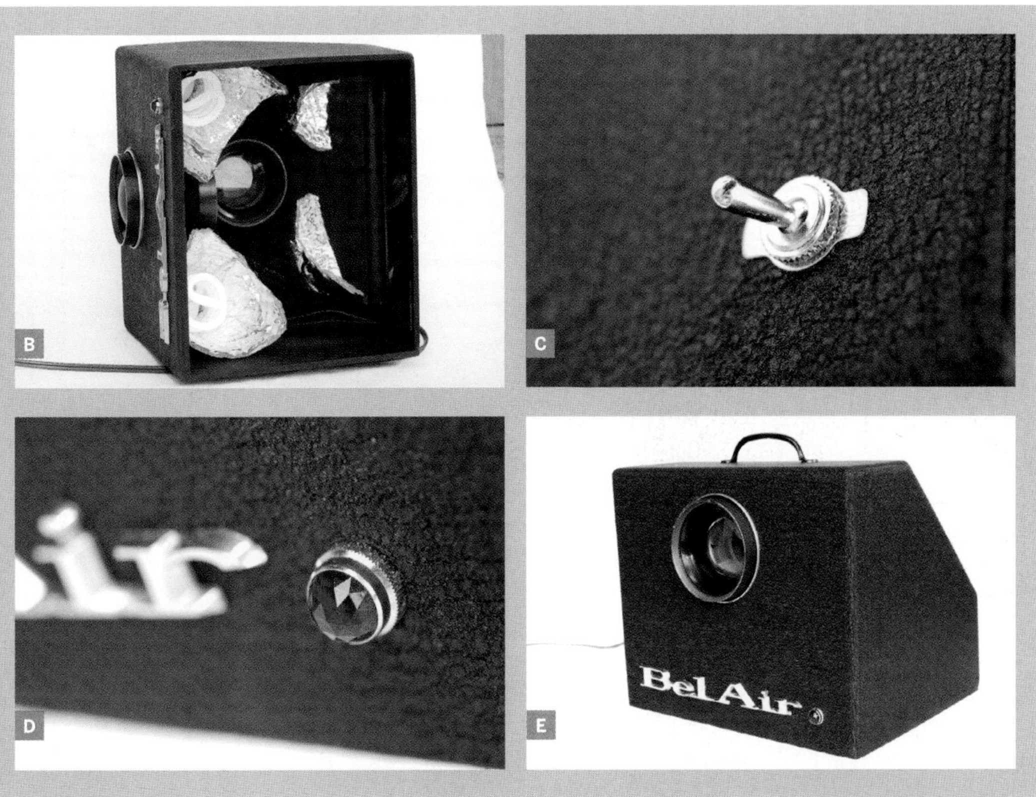

light bouncing around inside (Figure B).

Styling

I knew right away when I saw the quaint old Edmund pamphlet that I wanted to style my project such that at first glance, you might assume it was made long ago by your great uncle Lenny, the guy who was always tinkering in his machine shop, building his own riding lawnmower, or putting on a 3D slideshow of vacation shots from his Stereo Realist camera.

To evoke the stamped sheet metal construction of the old projectors that I admire, I routed the edges of the cabinet with a ½" round-over bit to soften the profile before applying a black wrinkle-paint finish. The wrinkle finish really gives the projector that old optical instrument look, like you find on ancient microscopes or vintage press camera accessories. It's become a favorite finish of mine as it looks really industrial and dries into an incredibly durable shell.

I've tried two brands of wrinkle paint, Plasti-Kote and VHT, and both worked well. To get the paint to wrinkle, you apply 3 heavy coats, waiting only long enough to allow a skin to form before applying the next coat. As the whole thing dries, it shrivels up to a nice, pleasing wrinkle. If it doesn't seem to be wrinkling after 1–2 hours, try placing in it direct sun or applying heat from a hair dryer.

I finished off the vintage look by adding a chrome toggle switch (Figure C), a "Bel Air" fender script from an old Chevrolet (Figure E), and a ruby-red power indicator jewel from a Fender guitar amp (Figure D).

I think Uncle Lenny would approve.

RESOURCES:

Edmund Scientific how-to pamphlet (PDF): makezine.com/go/opaque

American Science & Surplus: sciplus.com

Surplus Shed: surplusshed.com

Steve Lodefink is a designer by day, and by night likes to learn new skills. Building small projects using methods and materials that he's never tried is his hobby and his therapy.

SCANNER CAMERA

Mod a flatbed scanner to take photos that deconstruct time and motion.

By Mike Golembewski

A scanner's image sensor captures a scene slowly, line by line.

Several years ago, I built my first scanner camera. The idea was simple: I would use an ordinary flatbed scanner with a homemade large-format camera. The camera would focus the image onto the scanner bed in place of photo paper or film. I expected this to be a quick little art project made with a cardboard box, the cheapest flatbed scanner I could find, and lots of duct tape.

But when I got it all to work, the results were wonderful. Stationary objects photographed normally, but moving objects appeared twisted and distorted into fascinating shapes. At first I thought there was something wrong with my contraption, but then I realized that the movement of the scan head was meshing with the movement in the recorded scene. The distortion is similar to the effect created by moving an original on a photocopier mid-copy, but extended into the real world.

Making and using a scanner camera is a lot of fun as a technical exercise, but more importantly to me, it provides an interesting photographic perspective on time and movement.

Here's how to build 2 versions: a simple cardboard-and-duct-tape one that keeps the scanner intact, and a warranty-voider version that's more portable and flexible, and takes sharper pictures.

Photography by Mike Golembewski

SIMPLE SCANNER CAMERA PHOTOGRAPHS: Traffic study at Notting Hill Gate, London; Moreen and Rowan, Brighton; traffic study at Queensgate, London.

MATERIALS

SIMPLE SCANNER CAMERA

Canon CanoScan LiDE 20, LiDE 25, or LiDE 30
Computer
Black foamcore board, ¾" thick **I used Gatorboard.**
Cardboard, ¼" thick
Heavy cardstock
Tracing paper
Duct tape
Magnifying glass with ¼"-thick lens
Strong glue
Ruler
Hobby knife

DELUXE MODEL SCANNER CAMERA

Simple Scanner Camera **as listed above**
Mac computer running OS X
Black electrical tape
Duct tape
Velcro tape
Paper
X-Acto knife
Dremel tool with sanding and abrasive point bits
Small file
Fine sandpaper
Tweezers
Needlenose pliers

Simple Scanner Camera

1. Build the baseboard.

Cut a piece of black foamcore that fits exactly over your scanner's glass bed, then cut a 7" square hole out of the center. This will be the baseboard for your camera.

2. Make the boxes.

Make 2 boxes that slide together for focusing. Using cardboard and glue, make a 7"×7" inner box with both ends open, and then an outer box with a lid on top, slightly larger than the inner box, so that they nest snugly together. Line all box edges with duct tape. Cut a 3½"-diameter hole in the lid of the outer box.

3. Make the lens board and aperture cards.

Remove the lens from your magnifying glass and cut a hole in the center of a 6"×6" cardboard square to hold it. Tape the edges of the lens securely into place on the cardboard (Figure A). This is your lens board. Out of heavy cardstock, cut a set of covers for the lens, with different-sized holes in the middle. These are the aperture cards, which you'll tape

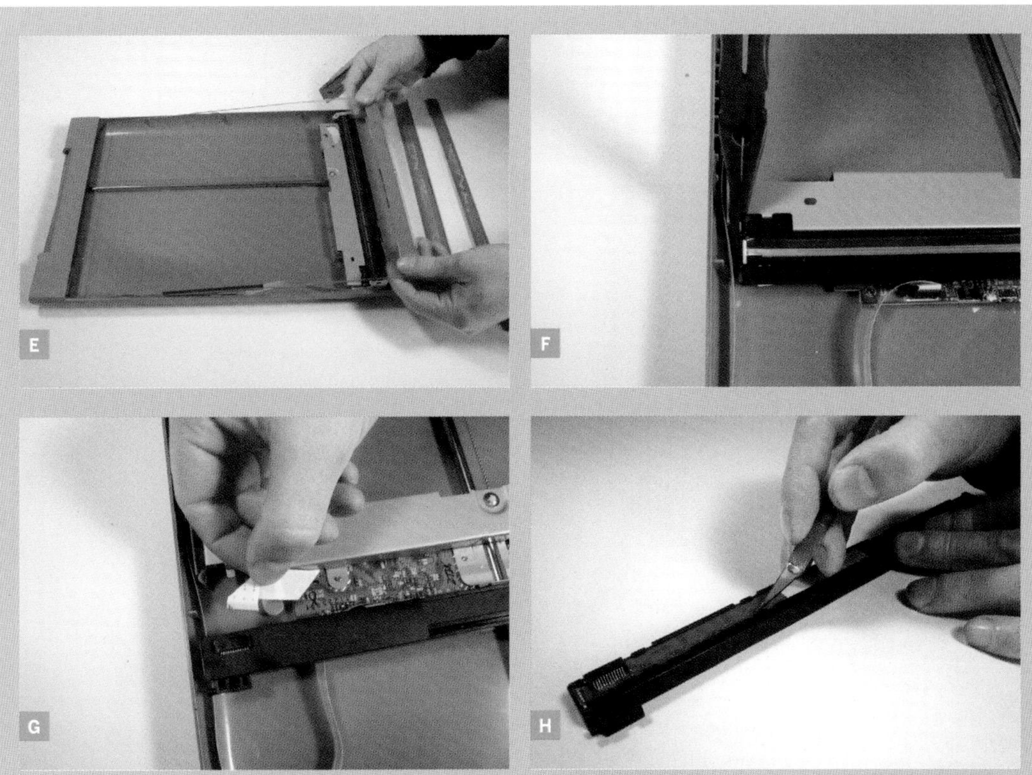
E
F
G
H

over the lens to control how much light gets into the camera, just like an iris in a regular camera (Figure B).

4. Assemble the camera.

Fit 1 end of the inner box into the baseboard and duct tape it in place from the inside (Figure C). Slip the outer box over the inner box and make sure you can slide it back and forth. Tape the lens board to the outer box with the lens centered over the 3½" hole.

5. Take some photos.

Your scanner camera is ready to go! To focus it, tape a piece of tracing paper over the hole at the back of the baseboard, then point the lens toward a brightly lit scene. Slide the outer box back and forth until the image comes into focus on the tracing paper. With my 2½" magnifying glass lens, I needed a focal distance (distance between lens and image) of about 7" to 12" for objects in the same room.

Tape the camera to the front of your scanner and start up your imaging application. Use the Preview button for fine-tuning the focus, and when you're ready, click Scan to take a picture. To adjust the image brightness, try different lens aperture cards (Figure D).

Deluxe Model Scanner Camera

Photographs taken with the simple scanner camera typically feature heavy vignetting, where brightness drops off farther away from the center. Also, the scanner's lamp can add undesired interference to a shot's lighting.

If you're willing to mod your scanner and dedicate it to camera use, you can get higher image quality and greater flexibility. Once you make these modifications you won't be able to use it as a normal scanner. And if you mess things up by working too quickly, you run the risk of rendering your scanner permanently and totally useless. Assuming you're still with me, read on.

1. Install the software.

First you need to install open source drivers from the SANE project (Scanner Access Now Easy) that allow your scanner to skip the calibration step and take scans even after being hacked. Download the

latest version of the TWAIN SANE interface at www.ellert.se/twain-sane and follow the installation instructions.

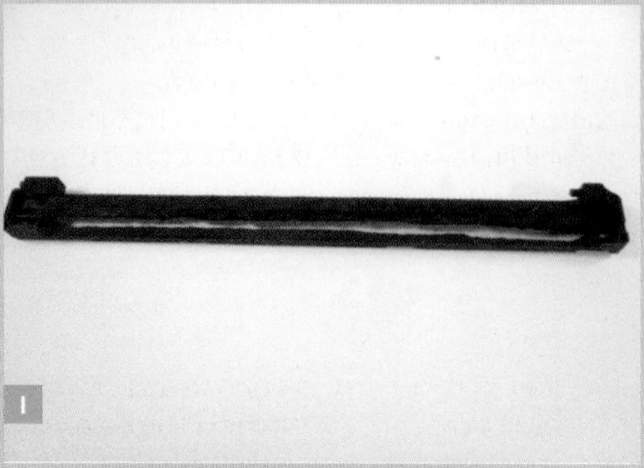

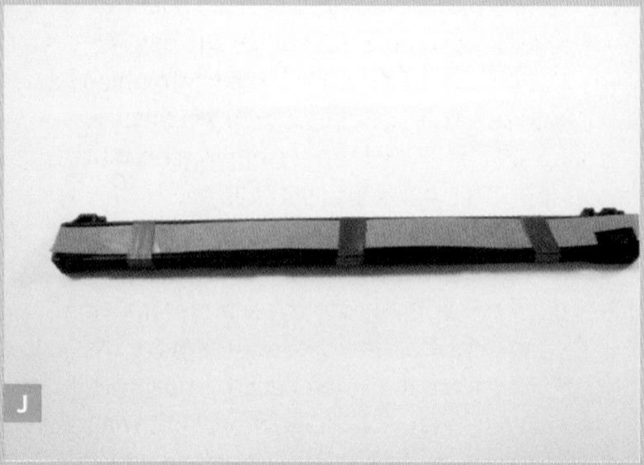

You should be able to access your scanner via the SANE-TWAIN plugin from any TWAIN-compliant imaging application.

With the software installed and the scanner plugged in, you should see a new SANE item in your Mac's System Preferences. Open it up, find "plustek" in the driver list, and click its Configure button. This will open the preferences file for the back end of your scanner, where we'll need to change a few lines:

On line 105, change `option skipCalibration 0` to `option skipCalibration 1`

On line 111, change `option skipFine 0` to `option skipFine 1`

On line 116, change `option skipFineWhite 0` to `option skipFineWhite 1`

Click OK to confirm the changes. Your scanner software is now ready to use with a modified scanner camera.

2. Open up the scanner.

Remove the lid from the scanner; you won't be needing it anymore. Two gray rails run along the long sides of the scanner and are held in place with tape. Insert the tip of your hobby knife under each rail and gently pry it up until it detaches. Remove the rails, then lift off the glass plate (Figure E, previous page). Set the glass and rails aside, and try not to touch the glass more than you need to.

3. Take apart the scan head assembly.

Locate the scan head assembly, which pulls itself back and forth on a geared thread. Gently pull it to the middle of the scanner, then orient the scanner so that the scan assembly runs left to right and the ribbon cable that feeds it bends toward you.

The assembly has 2 main components: a gray metal housing that contains motors and electronics, and the thinner, black plastic sensor bar with the image sensor and lamp. Find and remove the white plastic tabs at each end of the sensor bar and set them aside.

The sensor bar is secured to the scanner housing

by a metal clasp on its left side. Use pliers to bend the clasp so that it no longer holds the peg (Figure F, page 81). You'll need to bend it back into place later, so don't mangle it. There is also a small spring on the right side of the scanner housing, beneath the sensor bar. Take it out and set it aside.

4. Remove the sensor bar.

The sensor bar connects to the scanner housing by a white ribbon cable. Carefully remove the ribbon cable from the sensor bar by slowly pulling it straight out (Figure G, page 81). Don't wiggle it. Place the sensor bar on a clean surface and keep the rest of the scanner and the pieces you removed in a safe place.

5. Remove the contact image sensor (CIS).

From this point on, you need to be careful. The contact image sensor (CIS) is the green PCB inside the sensor bar. It's delicate, and damaging it renders your scanner useless. The CIS is also what lets us make this camera; it's easy to hack and uses very little power, allowing the scanner to be powered by a laptop USB connection.

Place the sensor bar facedown. Carefully remove the 3 pieces of tape covering the CIS and set them aside. Next, remove the 10 molded plastic tabs that hold the CIS in its black plastic housing. Use an X-Acto knife to lightly score along the base of each tab until it falls off (Figure H). Use tweezers to remove each detached tab. Make sure that neither the knife blade nor the tweezers touch the CIS. Take your time; this should take 20–30 minutes.

Remove the green PCB from the sensor bar. Use as little force as possible. If it won't lift out easily, work on the tabs some more, then try again. Once it's removed, place it gently in a clean, safe place, away from your work area.

6. Modify the sensor housing.

You need to allow as much light from the lens as possible to hit the CIS. To do this, you'll modify its plastic housing. First, carefully remove the clear plastic prism and the black rectangle of dense plastic that run the length of the sensor housing.

Now you'll want to cut out the middle and flatten the top of the sensor housing, so that more light can get through to the CIS and the sides won't cast shadows. Use the Dremel with a sanding band bit to grind the top of the sensor bar flat. Work at a slow speed — if you try to work too quickly, the plastic will melt. Once you've flattened down the top of the sensor bar housing, use the abrasive point tip to clear out a long slot in middle (Figure I).

Clean off excess shreds of plastic with your X-Acto knife and fine sandpaper. Vacuum everything up at this point — you don't want any of that black plastic dust getting onto the sensor or into the scanner.

7. Cover the lamp and replace the CIS.

Bring the CIS board back to your workspace. The white tab sticking up at 1 end is the LED light source. Cover it completely on both sides with electrical tape to prevent any light from escaping. Then insert the CIS back into the modified housing and replace the original pieces of black tape on the back of the bar. Cut a piece of paper to fit over the entire bottom of the sensor bar and tape it in place with small pieces of duct tape (Figure J).

8. Put the scanner back together.

Take the scanner back to your workspace. Reconnect the white ribbon cable to the sensor bar. Place the sensor bar into position, and bend the metal clasp back into place. Replace the spring and the white plastic clasps. Add 2 folded-over tabs of duct tape to the scanner assembly, next to the metal bar (Figure K). These will help prevent reflections from the metal bar from affecting your image.

Replace the glass plate — make sure it's clean first! Reattach the gray plastic strips. If you need to, use tape to secure them. There! Your scanner is back in one (modified) piece.

9. Attach the camera.

Use velcro tape to attach your camera to the scanner (Figure L). Attach the pointy side of the velcro to the glass (to avoid scratches). Run duct tape around the outside of the baseplate, to keep out all outside light.

Your camera is complete. Plug it into your computer, start up your imaging application, and load the SANE-TWAIN plugin. If everything went according to plan, you'll be taking scanner photos in no time. Follow the method described in Step 5 of the simple camera. Explore the options in the SANE-TWAIN interface; they give you a high level of control over the image quality. Enjoy, and good luck!

Tips for the Scanner Photographer

Once you've made the deluxe scanner camera, think of the scanner itself as a photo back, and try using it

with other camera bodies. I've assembled scanner cameras from Brownie box cameras, cardboard boxes, PVC tubes, and even old stage lights and magic lanterns. With a little effort, you can also mount a scanner camera back on a large-format monorail or field camera.

Large-format lenses for cameras are expensive, but you can find lenses that may work just as well inside old photo enlargers, overhead projectors, stage lights, and even toys!

The nice thing about the CIS-based scanners used for this project is that they're powered via USB. Hook your camera up to your laptop, and take it on the road.

The motion distortion is the most interesting thing about the scanner camera. Spend some time getting to understand it, and you'll start thinking about time and movement in photography in a whole new way — and this will start to inform your traditional photography as well.

For more photos taken with the simple and deluxe scanner cameras, see makezine.com/14/scannercamera.

Michael Golembewski is an artist and interaction designer who lives in Brooklyn, N.Y. Visit his Scanner Photography Project at golembewski.awardspace.com.

DELUXE SCANNER CAMERA PHOTOGRAPHS: Portrait of Abigail Durrant, London (opposite); traffic study with moving bus at Hyde Park, London; portrait of Astrid Askberger, London; in the pub at the All Tomorrow's Parties festival, Camber Sands, U.K.

Make Noise!

Turn your desktop computer into a musical instrument. By Charles Platt

Electronic music originated largely during the 1950s in the BBC Radiophonics Workshop, where reclusive "boffins" soldered resistors and vacuum tubes to create synthetic compositions. They recorded the output onto pieces of open-reel ¼" tape, which they spliced together using razor blades on editing blocks.

About 10 years later, John Lennon used tape loops to create the eerie repetitive patterns in the Beatles song "Tomorrow Never Knows." Ironically, now that we can synthesize sound without resorting to such drudgery, electronic music has sunk into relative obscurity (I'm talking about serious compositions rather than the simple riffs of techno). Still, an international community of sound synthesists exists online, and all you need to participate is the computer that you already own.

Below are 3 applications that can help you make electronic music. Audio is processor-intensive, so you'll be able to do more if you have powerful hardware — but if you just want to demo the software below, even the cheapest eMachine will do.

1. THE SOUNDRY

To learn the basics about sound and how we perceive it, see this excellent audio section in the Oracle ThinkQuest educational site at library.thinkquest.org/19537. Click the Wave Applet and you can draw your own sound wave in this beautifully implemented piece of Java (Figure A). After drawing a wave, listen to it while you refine it by adding attributes such as sine (for the melodious qualities of a sine wave) or sawtooth (for a fuzz effect). This is a great introduction to sound-wave fundamentals.

2. ROLLOSONIC

When something really scary was going to happen in a 1950s horror movie, most likely you heard a theremin on the soundtrack. This primitive electronic instrument, named after its Russian inventor, consisted of a box sprouting 2 antennas, one controlling pitch, the other controlling volume. To play it, the operator would make hand passes like a magician. Internal electronics detected the varying capacitance of the human body and emitted an eerie, quavering note through a loudspeaker.

For a modern, more full-featured, virtualized version of the theremin, go to rollosonic.com and click Download Now. Double-click the program icon, choose No for the installation option, and the software will run without needing to be installed. Click the Start button, click Get New Module, and select the Ding-Dong Module to create some noise. You'll see a floating menu with pull-down options, each allowing you to control a different aspect of the sound.

While the theremin responded to hand movements, RolloSonic responds to mouse movements. The position, direction, and even the acceleration of the mouse can be used. For a quick demo, set Horizontal Mouse Position as the Note-Input Source, leave Note-Velocity Control with a low Manual setting, and use Vertical Mouse Position for Note-Length Control (Figure B). Now slide your mouse around its pad and listen.

There's no way you're going to make beautiful music with RolloSonic. The screeching, buzzing, whistling, and burbling will be unpredictable, confusing, and often quite horrible. Still, it's unique, and you can use the program to develop, distort, or (with very little effort) destroy sound inputs from other sources such as a microphone. Not entirely practical, but fun, and free!

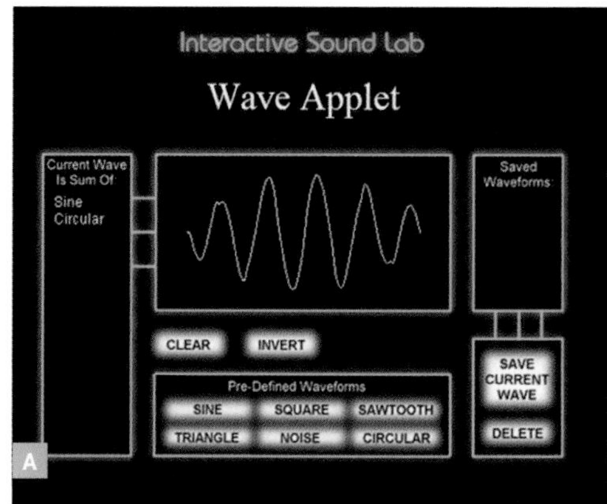

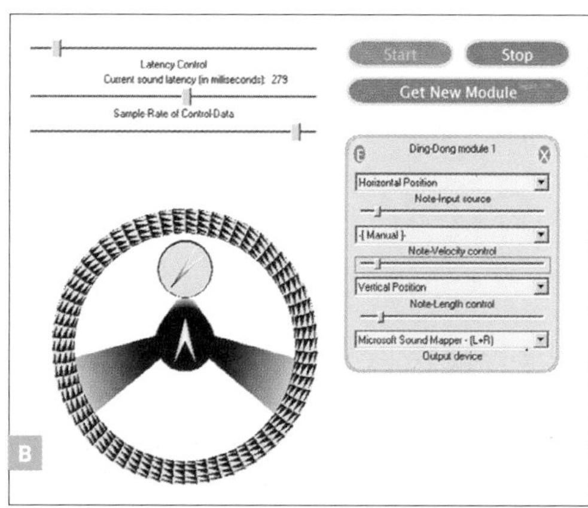

3. AUDIOMULCH

Enough foolishness. Time now for a serious and very powerful sound-synthesis application: AudioMulch, created by Ross Bencina, an Australian performer of electronic music. You can use it free for a 90-day evaluation period, or buy it for $89 — not unreasonable, considering that this is in many ways an audio equivalent of Photoshop.

For your free trial, download and install the software from audiomulch.com, launch it and decline the invitation to read help, then pull down the File menu and open a sample from the Examples folder. Click the green arrow in the task bar to play the audio, and I think you'll be impressed. You'll also see, on the screen, exactly how it was created. Sound origination and modification modules are chained in the Patchboard pane, and the properties of each module are shown adjacently (Figure C).

You can drag modules around, break and remake the links between them, twiddle knobs, and hear the results. To go beyond trial and error, open the excellent Help file and try the Beginners' Tutorial, which takes you through half-a-dozen simple steps to develop your own techno riff in a matter of minutes. Within a couple of hours I created my own composition (Figure D).

AudioMulch accepts VST plugins but is sufficient on its own to function as a serious musical instrument. You can save your sounds as a *.wav* file, or play them in real time. The results can be as melodic or as synthetic as you wish.

Charles Platt (platt@makezine.com) is the editor for MAKE's Upload section.

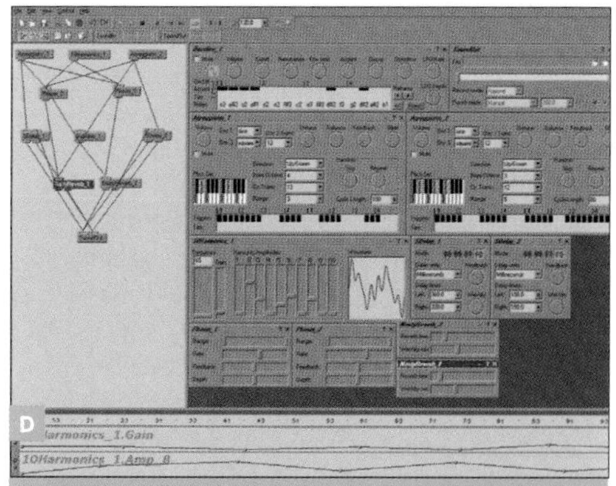

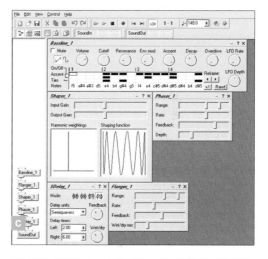

Fig. A: Sine wave modulated by circular wave, sounds good! **Fig. B:** Mouse-controlled noise, sounds not so good. **Figs. C and D:** The Patchboard interface on AudioMulch creates a melodic audio environment.

Page Yourself

When you want to put up your own website, how free can a freebie be?
By Brian O'Heir

Let's face it, if you Google yourself and find only hits for other people who have the same name, you're not gonna get that special feeling — that feeling of purpose, place, and accomplishment in society.

I wanted to be a search-engine somebody, with my own personal website to represent me. Also, I wanted to share events and photos with friends and relatives. And naturally, I didn't want to pay for it!

Initially I searched for "free websites," but the ones I found were hardly free. Intuit is free for 1 month, and $5 per month thereafter. Microsoft is free for 1 year, but then you pay even more. The Yahoo! Geocities Page has the word "free" all over it, but nothing can be done without paying money. Freewebs.com is genuinely free, but is ad-supported, and really slow, and it crashed my browser. That kind of "free" costs too much.

One genuinely free option is Google Page Creator. It's a Google Labs beta product, which may not work perfectly and may change as refinements are added. Probably the basic version will remain free when more powerful versions become available for license.

Currently, each Page Creator account allows you three websites with a total size of up to 100MB. Best of all, the system is extremely easy to use, and should enable literally everyone who can use a computer to build their own website.

1. SIGN UP

Go to pages.google.com and you'll be asked to give your Gmail address. If you don't have one, click the link at the right to create one. Your Gmail username will then be used to create the URL for your home page in the format *http://yourgmailname.googlepages.com*.

(You can also create and register your own URL, and point it to your Google site, but we'll get to that later.)

2. ADD TEXT

Click "I'm ready to create my pages" and accept the inevitable terms and conditions. You'll now find yourself in Page Creator's Layout Manager (Figure A).Make up a title for your page (you can change it later) and an optional subtitle. I used "Ball Toy" because I wanted to promote a toy that I designed for children and cats. Now type text in the main content section, or just copy and paste it from another window or application. Note the button bar, which allows you to format any highlighted text, including the Heading button, which lets you apply preset text formats.

3. ADD IMAGES

Click the Image button, and the Add an Image window opens. Browse and open any image file on your hard drive, and a thumbnail of it will be added to this window. Everything you upload accumulates here. Next and Prev buttons navigate through groups of eight images, with a current limit of 500 files and maximum single file size of 10MB. Click the image you want, and then the Add Image button, to place the image on your page.

4. ADJUST IMAGES

When you drag an image, text will flow around it accordingly. This feature didn't work great and was a bit frustrating at times, particularly if the field contained no text. Temporarily placing some text to buffer images helped. Selecting any image on the page presents an Edit Image button, which opens a window where you can change size, brightness, contrast, and other features.

Google's Page Creator lets literally everyone who can use a computer build their own website.

5. ADD LINKS

A link is a small text section that will jump users to someplace else when they click on it. Select some text as the link, then click the Link button and choose the destination that the user will jump to: Your Pages (another page on your website), Your Files, an external Web Address, or an Email Address (Figure A). Once a link has been created it can be cut, copied, and pasted like any normal text.

6. STYLE THE PAGE

The Change Look button at the top right corner of Page Creator opens a set of 41 color and graphic combinations (Figure B), while the Change Layout button offers four different column configurations (Figure C). Try different combinations and see finished browser pages by clicking the Preview link at the top left corner (Figure D, following page).

7. EXTRAS

At the bottom right corner of the Page Creator window you'll find two links: one allowing you to edit the HTML code manually for each section of the page, the other to add "gadgets," which are little HTML programs that you can paste in from elsewhere.

Page Creator is quite limited, but this can be beneficial. With all the bells and whistles on the web today, one way to stand out is to not have all these effects on your pages. Keeping your pages simple allows you to concentrate on the content, rather than features. Clean copy, good pictures, and open layout will distinguish your pages from the congested, code-generated pages out there.

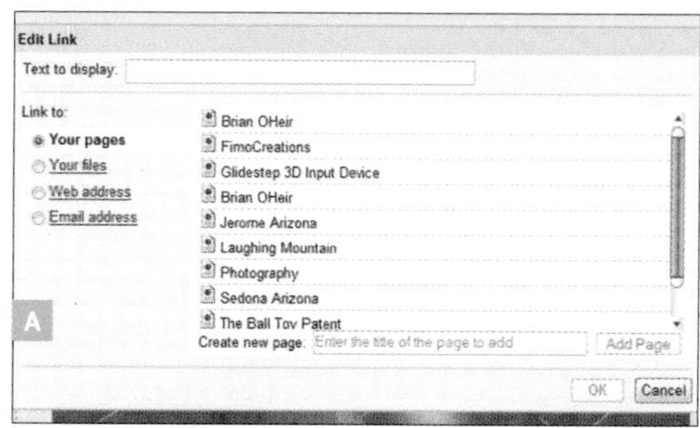

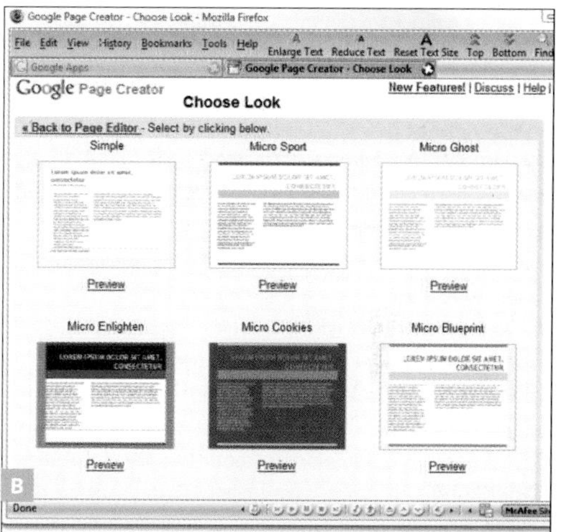

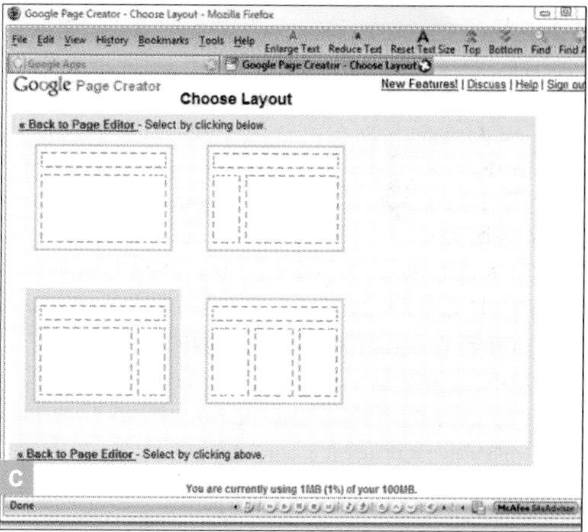

Fig. A: The link destination window, showing other pages on the site to link to. **Fig. B:** The Choose Look page, showing six of the 41 available page styles. **Fig. C:** The Choose Layout page, showing the four available layouts.

8. ADD MORE PAGES

Click the Back to Site Manager link, and the Site Manager shows you the pages already on your site, with the option to add more (Figure F). After you add them, Site Manager lets you jump from one to the next by clicking its icon. This is how you navigate through your site while you're building or maintaining it.

9. MAINTAIN AN INVENTORY

After a while you can lose track of all the text and images you've uploaded. The list of Uploaded Stuff on the right of the Site Manager window contains all the files you've uploaded. Clicking on any file opens it in a new browser window, and clicking on the trash icon deletes the file.

10. PUBLISH!

In the Site Manager window, click the checkbox on any page, then click the Publish button, and that page goes online. Using the More Actions drop-down menu you can also Tell your friends, Discard unpublished changes, Unpublish, Duplicate, and Delete. You can even hide your site, which will cause users to receive an error message when they enter its URL in a browser.

After I created my site, I decided I wanted to have my own domain name that would link to it. Luckily brianoheir.com was available so I bought it from the Yahoo domain service for $10. Afterward, at Yahoo's Domain Control Panel ⇒ Forward Domain ⇒ Create Forward page, any website can be specified as the destination for any domain name. I then forwarded brianoheir.com to brianoheir.googlepages.com.

My site brianoheir.com has been fun to make, and I like the idea that it represents me on the internet with very low maintenance. Of course, nobody visits it, so I haven't exactly become a star, and while it was the first hit for the search term "Brian O'Heir" on Google for several weeks, it has now been overtaken by another listing. But, I think it's OK — that new listing is "makezine.com: Brian O'Heir"!

Brian O'Heir is an architect and toy inventor who lives near Sedona, Ariz.

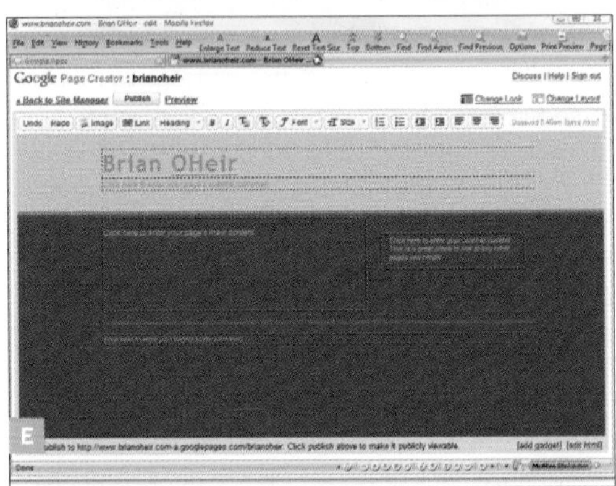

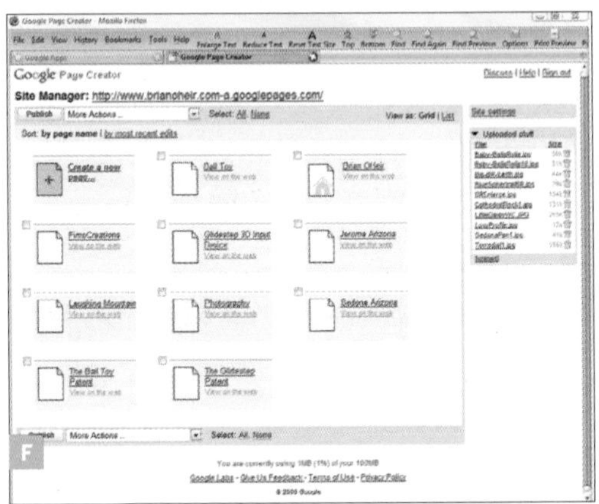

Fig. D: Preview of a web page, showing title, subtitle, body, and image. **Fig. E:** Layout Manager for the brianoheir.com home page, showing text button bar, Change Look and Change Layout links, and Publish button. **Fig. F:** The Site Manager window, showing icons for each page and the Upload stuff list (right).

As Good as Old

Make digital photographs look as if your great-grandfather took them.
By Richard Kadrey

Once upon a time, photographers took pictures on a delicate, wonky medium called "film," which was plagued with defects such as dust, scratched negatives, grain, and colors that faded with time.

USE DIGITAL TO CLICK BACK THE CLOCK

Digital photography eliminated those defects, yet somehow the photographs didn't quite look like photographs anymore. I found that I missed many of the dirty, human elements of film, so I ended up putting them back into my squeaky-clean digital shots. Here are some tips on how you, too, can ruin, damage, and age modern pictures that look too good to be true.

Your primary tool will be a program such as Photoshop or GIMP, either of which allows you to use layers. (GIMP is a powerful and free Photoshop substitute available for download from gimp.org.) Layers are stacks of images that can be superimposed like old-fashioned transparencies. You blend them to create something weird and new.

1. COLLECT TEXTURES TO ADD TO YOUR PHOTO

I've been shooting cracked concrete, stained floors, peeling paint, wood, and rusting metal for over 10 years, and have an extensive library I can dip into. If you don't have your own textures, you can find them free at sites such as deviantart.com, or you can buy high-quality images cheaply from a stock photo company such as iStockphoto (istockphoto.com).

2. SELECT A SUITABLE PHOTO

I decided to work with a shot I took of a model at a San Francisco train station. I wanted an image that looked something like a photographic version of one of Paul Delvaux's surrealist sleepwalker paintings.

Fig. A: The final photo, after all the tweaks and modifications. **Fig. B:** The original photo, as taken by the author.

Photography by Richard Kadrey

C

D

E

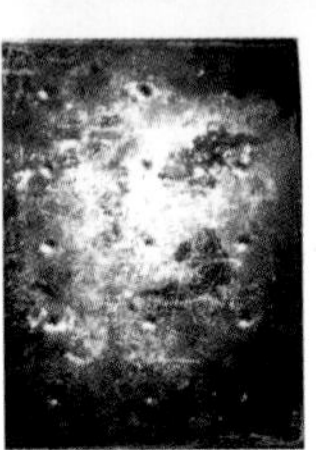

3. CONVERT TO MONOCHROME

I prefer to age black-and-white images because for me, it's all about the textures, and I don't want to worry about how colors might or might not blend together.

4. CHOOSE SOME APPROPRIATE TEXTURES

Do you want a photo that looks just a little worn and faded, or do you want something that looks like it fell down a mine shaft 100 years ago? I've obtained good results by superimposing textures of distressed paper, peeling paint, and the kind of wall and floor stains that you often find in old warehouses or industrial sites (see corresponding textures for Figures E and F).

5. ADD THE TEXTURES IN PHOTOSHOP

Open each texture photo, choose Select ⇒ All, then Edit ⇒ Copy, then go back to your primary photo and Edit ⇒ Paste. Photoshop automatically adds the texture as a new layer. I used 7 layers to create this sleepwalker shot (only 5 are shown here).

6. SUPERIMPOSE THE LAYERS

To control the way the layers superimpose, look for the Normal button in the Layers palette. Normal is the default mode for each layer. Click on it and hold it down, and you'll see a list of other options that determine how each layer will interact with the one below. For the example here, I used the Lighten, Overlay, Soft Light, and Hard Light tools.

This is mostly a matter of trial and error, but here are two tips:

» Try using Photoshop's Image ⇒ Invert command to reverse the colors in your textures, so that black cracks on a white wall become white cracks on a black wall. This inverted version will blend differently with the other layers.

» Use Image ⇒ Adjust ⇒ Brightness/Contrast to alter the lights and darks of your textures. Push the contrast way beyond anything you'd do to a regular photo. A layer that's only stark black and white with no grays can create dramatic changes to the underlying photo.

F

G

You, too, can ruin, damage, and age modern pictures that look too good to be true.

7. FLATTEN AND MERGE

When your image looks the way you want it, save a copy of the original with all the separate layers intact. Then use Layers ⇒ Flatten Image to merge all the layers and the original photo into a single image. Save this as a separate copy. You can now make other changes to the flattened file.

In my sleepwalker image, I applied 4 layers of textures to the original image, then flattened them and dropped that image onto a stock photo of distressed paper that I found on iStockphoto. The black-and-white sleepwalker layer is applied to the torn paper as Hard Light at an 84% Fill.

I then went around the edges of the sleepwalker layer and erased some of the edges to match the torn and missing edges of the stock photo. I also erased some middle portions of the image to match the tears and cracks.

8. CONSIDER ADDING A LITTLE COLOR

This will help to create a different mood. Adding a sepia tone will make the image feel as if it's old, perhaps from the 19th century. Adding a hint of cyan will give the image a bright, silvery look. I gave my sleepwalker a sepia look, then went back to the torn paper layer, desaturated the color in the torn areas, and whitened them to create extra contrast within the dark image.

I have one last suggestion, and it might be the most important idea that I've mentioned. *Be open to mistakes*. Some of my favorites of my own images came about because I accidentally set a layer to Hard Light instead of Overlay or vice versa. Remember musician Brian Eno's terrific advice: "Consider your mistake as a hidden intention."

TIP: Create a textures library by taking photos of rust, broken sidewalks and windows, etc., as you go about your day. You can also buy stock textures from sites like istockphoto.com.

Richard Kadrey is a novelist and an exhibited photographer.

Make: Projects

Hungry for hot projects? Indulge the dune racer in you from the comfort of your couch, using live video to steer an R/C buggy. Satisfy your taste for chocolate chews and multidimensional math with a homebrew taffy pulling machine. Then for dessert, make a music visualizer that hits the sweet spot between retro and futuristic.

LIVING ROOM BAJA BUGGIES

By John Mouton

BUG'S-EYE VIEW

With wireless cameras on board, these radio-controlled racers give you virtual reality telepresence.

Do you like radio-controlled (R/C) cars? Do you like the desert, but hate the heat? Well, sit down and kick back as you engage in the excitement of Living Room Baja Buggy Racing. This off-road competition combines the fun of homemade R/C cars with the air-conditioned convenience of a fake, indoor desert landscape — without the big dollar price. There are no rules, no expensive automotive racing equipment, and a total disregard for public safety (because these cars are only 6" long and 4" tall).

You don't even have to look at your car as you negotiate the terrain — an onboard camera and a virtual reality headset can be installed for your lackadaisical safety!

So chase the dog, race against the kids, or just put the electrons to the PC board and hold on for a bug's-eye view of the whole racecourse. It's the best thing to hit urban living since the Man Cave.

FULL DISCLOSURE: The author is an employee of Microchip Technology, which manufactures some of the components used in this project.

Photograph by Robyn Twomey

Set up: p.99 **Make it:** p.100 **Use it:** p.105

John Mouton is an applications engineer with Microchip Technology's Security, Microcontroller, and Technology Development Division (SMTD).

R/C COMMAND, MICRO CONTROL, AND MINI VIRTUAL REALITY

In this project, we'll build an R/C Baja buggy that you can drive around the house using a pair of virtual reality display goggles that give you a view of the world as if you were literally in the miniature driver's seat. All of this is controlled by the user (you) via a hand-held radio transmitter and a wireless camera that displays what the buggy sees through the VR goggles.

I started by designing a simple motor control circuit, a voltage regulator circuit, and a PIC12F683 MCU. This gave me a small drive circuit that would move the buggy forward and in reverse, electrically brake and stop the buggy, and provide variable speed control. I sent my schematic diagram and Gerber files to a vendor and got back a printed circuit board, ready to plug my components into.

After the circuit was built, I bought parts on the internet to build the buggy. When I put the buggy into action I found that it performed well — in fact, better than I expected.

I added a wireless camera and virtual reality goggles. Now I drive the buggy all over the house without leaving the comfort of my couch. You don't have to add the wireless camera and virtual reality goggles to have fun with this project. In fact, the buggy will be less expensive and lighter without them. Nevertheless, this article provides step-by-step instructions on how to build an R/C Baja buggy with the whole enchilada.

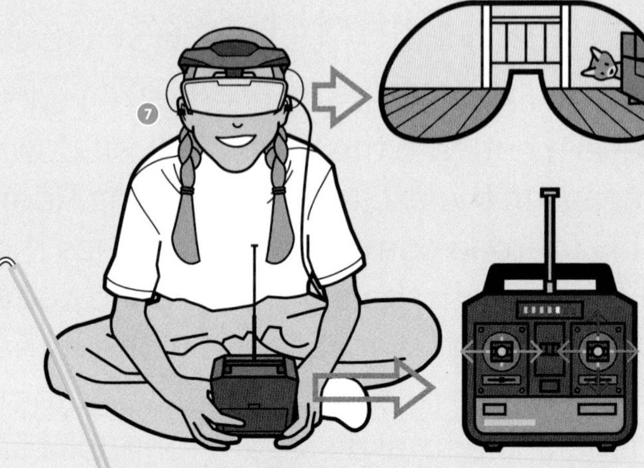

HOW IT WORKS:

1. The Baja buggy is built on a Tamiya Buggy Car Chassis Set.
2. The motor control circuit is built on a custom printed circuit board.
3. The drive motor (red arrows) makes the buggy go forward and backward.
4. A 9V battery and a lithium battery add weight to the back for improved traction.
5. One servomotor (blue arrows) pans the webcam mounted on top.
6. Another servomotor (orange arrows) steers the front wheels.
7. The wireless webcam signal is sent to the VGA virtual reality goggles.

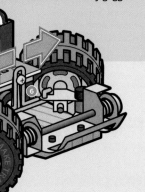

SET UP.

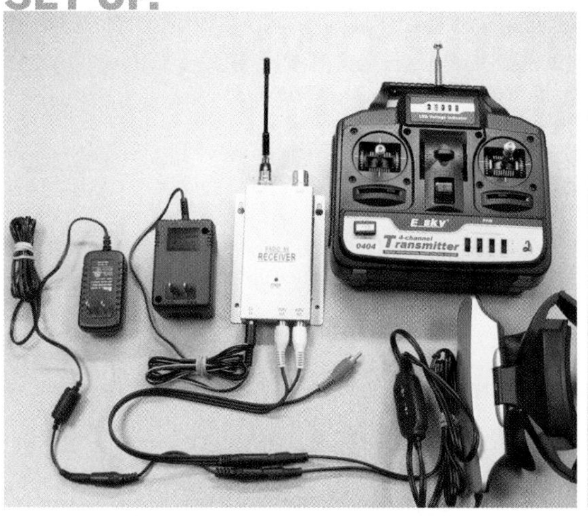

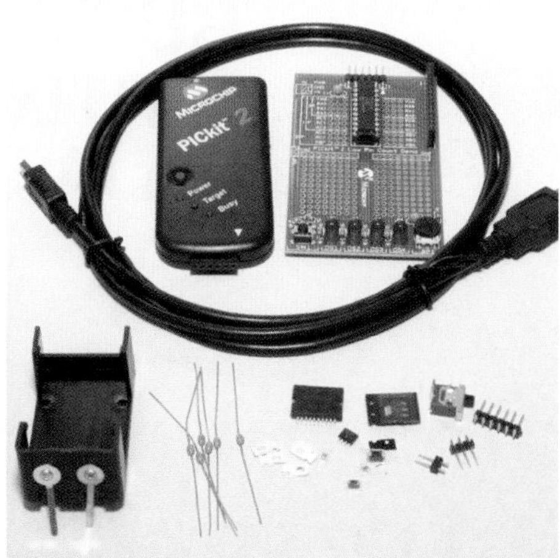

MATERIALS

Wireless camera and tuner/receiver **You need a wireless 9V DC, 300mA A/V security camera, and a 9V DC, 500mA tuner/receiver with 0.9G frequency control. Try** tigerdirect.com **or** outpost.com**, which are web warehouses with variable stock.**

VGA goggles **Any wireless-capable generic or name brand; try** amazon.com **or** tigerdirect.com**.**

Tamiya Buggy Car Chassis Set **#70112** tamiyausa.com

Tamiya Atomic-Tuned DC Motor **#15215**

E-Sky EK2-1003 4-channel radio system **which includes transmitter, receiver, and 1 servomotor (mini), or similar radio system**

E-Sky EK2-0500 servomotor **for the camera**

Futaba FM75.470 MHz CH.64 crystal **For multiple buggies, use different crystals in the 75MHz–76MHz range to avoid cross-talk.**

Thunder Power lithium polymer (Li-Poly) 1,320mAh/7.4V, 2-cell rechargeable battery **for the motor**

9V battery **for the camera**

Apache Smart Charger 2020 **for charging the Li-Poly batteries**

ElectriFly 2-pin male connectors **#GPMM3106, for all battery, camera, and charging connections**

Servo cable with 1 male S connector towerhobbies.com **#LXPWB5**

Male S connector **for servo cable, Tower #LXPWC3**

Ferrite ring **for radio-control applications,** deeteeenterprises.com **#K10031TA**

Battery holder digikey.com **#1294K-ND**

#4⅜" Phillips head screw, #4 flat washers (2), and 4-40 steel hex nuts (2) **for tie rod hardware**

NOTE: **For the R/C transmitter, receiver, servos, and crystal, try** hobbylobby.com **and** robotstore.com. **For the batteries and charger, try** rctoys.com **and** electrifly.com.

TOOLS

PICkit 2 Starter Kit **for programming** microchipdirect.com

Electric drill, cordless or AC

Drill bits **to fit the rivet diameter**

Center punch

Soldering iron and solder

Double-sided tape

Sandpaper

Computer

Phillips screwdrivers **large and small**

Side cutters, modeling knife, and scissors

Pop rivet tool, hand-held

Heat-shrink tubing

Heat gun or hair dryer

Wire strippers

Needlenose pliers

Fake desert landscape

DOWNLOADS

Download the Baja buggy motor control program for your PIC microcontroller, ***carprogorig.hex***, as well as the Gerber files. Then download a complete list of the circuit components. makezine.com/14/bajabuggy

NOTE: The whole project ran me $400, but if I didn't install the VR goggles and camera, it would total only $200: about $145 for radio-control components, $30 for the PCB and circuit components, and $20 for the buggy chassis set with motor. (I did order 10 PCBs for a unit price of $12. Ordering 1 PCB will run about $50.)

Photography by John Mouton (left) and Ed Iroxell

BUILD YOUR OWN VIRTUAL REALITY BAJA BUGGY

START Time: **4–5 Hours** Complexity: **Medium**

1. BUILD THE BUGGY CHASSIS

1a. Assemble the buggy chassis set as directed in part 3 of the instructions enclosed in the Tamiya Buggy Car Chassis Set box.

1b. Drill a hole in the back of the motor/gear bracket housing. You'll use it to rivet on the 9V battery holder. Use the same drill bit size as the size of your rivet.

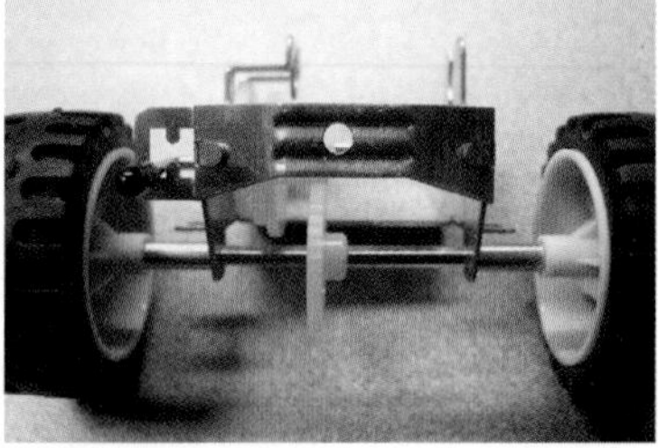

NOTE: Center-punch the hole before you drill, and use a high-quality drill bit with a sharp tip. This will prevent the drill bit from skipping across the metal and biting in the wrong place.

1c. Modify the tie rod. Bend and trim a 16-gauge wire to about 1" high with arms that fit against the tie rod provided in the buggy car chassis kit, as shown. Then solder the 2 together. This will give the servo a way to move the tie rod left, right, up, and down with the suspension, thus steering the buggy.

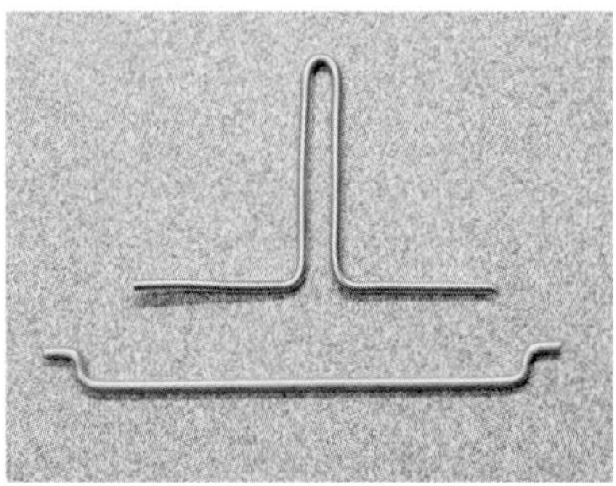

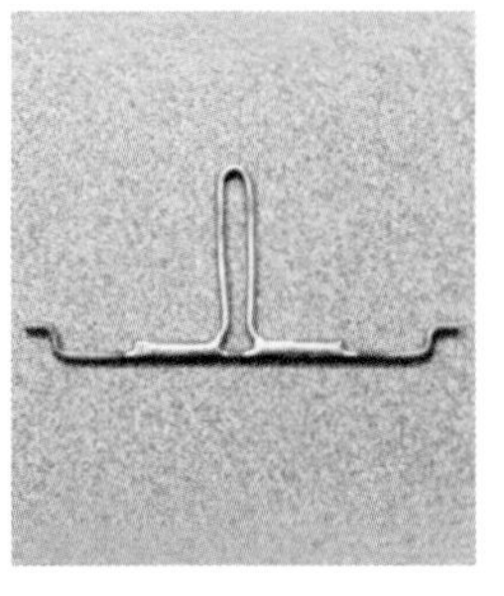

Photography by John Mouton

1d. Solder one 0.1-microfarad (µf) capacitor and a 24-gauge wire from each lead of the atomic-tuned DC motor to the motor case. Next, solder a third 0.1µf capacitor between the leads. This will prevent electrical motor noise from interfering with the receiver circuit.

1e. Install the wheels as directed in part 6 of the buggy car instructions. However, don't install the roll bar — it will just be in the way later when you install the camera servo and the motor-control circuit board.

2. BUILD THE MOTOR-CONTROL CIRCUIT

2a. Get a printed circuit board. I emailed the schematic and Gerber files provided at makezine.com/14/bajabuggy to Circuit Express (circuitexpress.com). I chose their least expensive package, 10 boards for $120, and they built the boards and mailed them to me within 24 hours. Other vendors are less expensive for smaller quantities. Or you can buy a PCB build kit and make it yourself.

2b. Install the circuit components. All components used in the motor-control circuit can be purchased from Digi-Key (digikey.com) or any other electronic components supplier. For the battery, camera, and charger connections, I used ElectriFly 2-pin male connectors to ensure safe, secure connections. I chose the PIC12F683 MCU because it's small, has a pulse-width modulation (PWM) module onboard, and very simple to work with. Here's what the completed circuit board should look like after all the components are soldered. Use the schematic to guide you. A list of components and part numbers is available at makezine.com/14/bajabuggy.

3. INSTALL CIRCUIT BOARD, BATTERY HOLDER, AND SERVOMOTORS

3a. Before you solder the motor leads, twist them together. This will keep electrical motor noise from interfering with the receiver. Now solder them to the circuit board at the locations marked M1 and M2.

3b. Use a small piece of double-sided tape to attach the circuit board to the wooden chassis, all the way back so that it touches the motor/gear bracket housing.

3c. To install the 9V battery holder, first sand off about 1/16" of its bottom end. This ensures that the battery will fit flush inside, and that the holder itself will fit evenly and spaciously between the rear wheels.

3d. Drill a hole in the center of the battery holder the same diameter as the hole you drilled in Step 1b. Then rivet the battery holder to the motor/gear bracket housing.

NOTE: To rivet, insert a rivet through the hole, and then squeeze with the pop rivet tool until the shaft pops off. To remove a rivet, drill it out using the same bit you used to drill the original hole.

3e. Extend the battery holder's positive and negative leads by soldering 24-gauge wires to them, long enough to comfortably reach J4 on the motor-control circuit board. Solder a 2-pin male connector to the ends of these wires, which will allow you to disconnect the 9V camera battery from the circuit board as needed.

3f. Using a small piece of double-sided tape, attach 1 servomotor to the motor-control circuit board as close to the motor/gear bracket housing as possible. This servo will be used to pan the wireless camera that will be mounted on top of it.

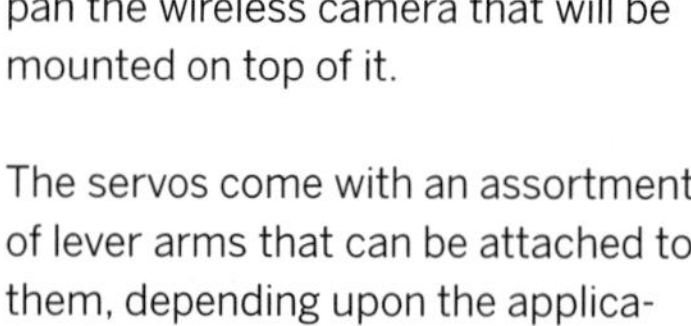

The servos come with an assortment of lever arms that can be attached to them, depending upon the application. Use the circular lever arm on the camera servo.

3g. Next, attach the other servo to the chassis just in front of the circuit board (the location is shown in Step 3e). This second servo will be used to steer the buggy.

3h. For the steering servo, use the straight lever arm. Drill a small hole in the lever arm the diameter of a #4 screw. Use the #4 screw, 2 washers, and two #4 nuts to loosely attach your modified tie rod to the servo lever arm. Now, as you drive the car over rough terrain, the servo will still be able to steer the front wheel even as the front suspension moves up and down.

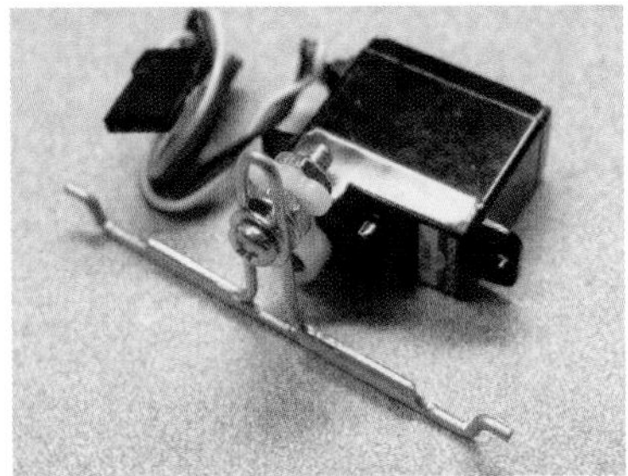

4. INSTALL THE R/C RECEIVER AND WIRELESS CAMERA

4a. You can use just about any R/C transmitter/receiver set for this project. If you don't already have a set, shop around for a good deal. I found a great deal on a 4-channel radio system kit that included the transmitter, receiver, and 1 servo for $60.

The reason I chose a 4-channel dual stick transmitter (rather than a 2-channel pistol-type transmitter) is because I needed the second left-to-right stick to pan the camera.

4b. Using double-sided tape, attach the receiver to the top of the steering servo. To reduce the receiver's bulkiness, remove its cover. This allows you to easily install frequency crystals and plug components into the receiver.

4c. Plug the camera servo into channel 4, and the steering servo into channel 1. Plug connection JP1 of the motor-control circuit board into channel 3 of the receiver; this connection will supply power to the receiver and allow you to control the speed of the motor via the PWM signal generated from the PIC12F683 MCU. Make the PWM signal cable by crimping the male S connector's pins to the bare-wire end of the male servo cable using a pair of pliers. Now, in order to reduce electronic noise interference, loop the PWM cable through a ferrite ring before plugging it into channel 3 of the receiver. This ring will increase the cable's inductance, thereby filtering out high-frequency electronic noise.

CAUTION: When using R/C transmitters and receivers for cars, keep them in the R/C car frequency range of right around 75MHz. R/C airplanes operate right around the 72MHz frequency range.

4d. Before installing the wireless camera to the camera servo, shorten the camera's cable and solder another 2-pin male connector so it will plug into the motor-control circuit board at J3. To do this, cut the camera's cable down to about 3". Strip 1" of insulation and slide on a 2" piece of heat-shrink tubing. Then solder the black and red wires to the 2-pin connector, slide the heat-shrink tubing over the connection, and shrink it with a heat gun or hair dryer.

4e. Using your double-sided tape, attach the wireless camera to the camera servo temporarily (you'll adjust it later).

5. INSTALL THE MOTOR BATTERY AND PROGRAM THE PIC MICROCONTROLLER

5a. For the motor's power source, you need something that will supply enough voltage and last a while between charges. I used a Thunder Power lithium polymer (Li-Poly) 1,320mAh/7.4V 2-cell rechargeable battery. I don't recommend using less than a 480mAh/7.4V battery, as it would need to be recharged more often.

CAUTION: The battery will have a charge even if you buy it brand-new, so keep the 2 wires from touching!

5b. Strip the black and red wires coming off the battery. Solder the wires to another 2-pin male connector, and heat-shrink. Attach the battery to the 9V battery holder with double-sided tape, and connect the rechargeable's wires to J5 (VBAT) on the circuit board.

5c. To program the PIC MCU, I used a PICkit 2 Starter Kit (#DV164120) that I bought from my company, Microchip Technology (microchipdirect.com) for $50. This kit has everything you need to write, debug, and program your source code directly into the MCU via header J2 on the circuit board. (If you're a student or educator, visit microchip.com/academic to learn how you can get a discount on development tools through Microchip's Academic Program.) Go to makezine.com/14/bajabuggy. All you have to do is to import the *carprogorig.hex* program file using the PICkit 2 Starter Kit interface, and then click the Program button.

FINISH X

NOW GO USE IT »

USE IT.

CHARGE IT UP

CHARGE IT

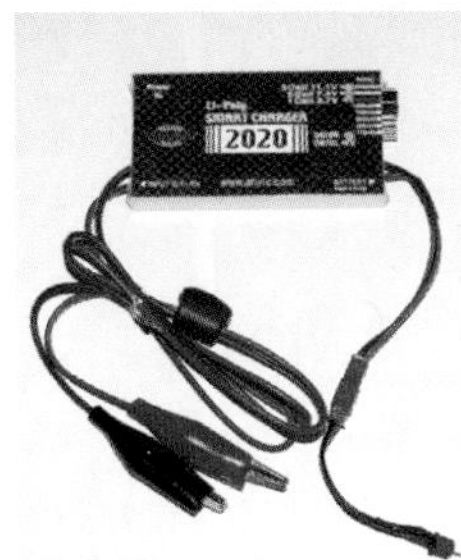

Now that your buggy is assembled, you're almost ready to turn it on. However, you might want to charge the battery first. I used an Apache Smart Charger 2020 designed specifically for charging 2-cell Li-Poly batteries.

CAUTION: Trust me when I say that if you use lithium polymer batteries, you *must* use a charger designed to recharge Li-Poly batteries. I have a coworker who no longer has a garage because her husband didn't use the appropriate battery charger. Always follow the manufacturer's instructions on proper use of batteries and chargers.

ZERO THE SERVOS AND SET THE RADIO CONTROLS

You're now ready to fire up your buggy. Once you turn on S1 on the circuit board, the servos will power up and find their neutral positions. Hence, you'll need to detach the camera and reattach it so it faces toward the front of the car. (Good thing you used double-sided tape!) You may also need to make a slight adjustment to the steering servo. Simply remove the screw holding the lever arm to the servomotor, pull it off, and reattach it, pointing straight up. Now, when you turn S1 off and then turn it back on, the servos will be aligned to the correct starting points.

With the circuit board and hand-held transmitter both turned off, adjust the transmitter toggle switches in the lower right-hand corner to set the left control stick (forward and back) as the throttle; the left control stick (left and right) for steering; and the right control stick (left and right) for panning the camera. The radio system kit's instruction manual will help you do this.

When finished, turn the transmitter on, then turn the circuit board power on at S1. Take the throttle control stick on the transmitter and pull it all the way back toward you. Then push it all the way forward, then back to center. This will establish the maximum duty cycles for forward and reverse, and arm the speed control circuit. At this point, the buggy is ready to roll.

USING THE VIRTUAL REALITY GOGGLES

Plug the VR goggles (or a TV monitor) into the receiver that came with the wireless camera. Plug the receiver into an AC outlet. Adjust the tuning knob until a picture of what the camera sees comes up. I found that the wireless camera range is more than 500' straight line-of-sight, with no major obstacles. If you use it outside in a large open area, the range is even better.

RESOURCES

Download the Baja buggy motor control program for your PIC microcontroller, *carprogorig.hex*, as well as the Gerber files and circuit schematic from makezine.com/14/bajabuggies. You can also download an Excel file listing all the Digi-Key part numbers for components that fit this circuit board. And finally, there's a list of lower-cost build alternatives provided.

Microchip's Online Motor Control Design Center (microchip.com/motor) has a wealth of resources for programming your PIC microcontroller to control motors and more.

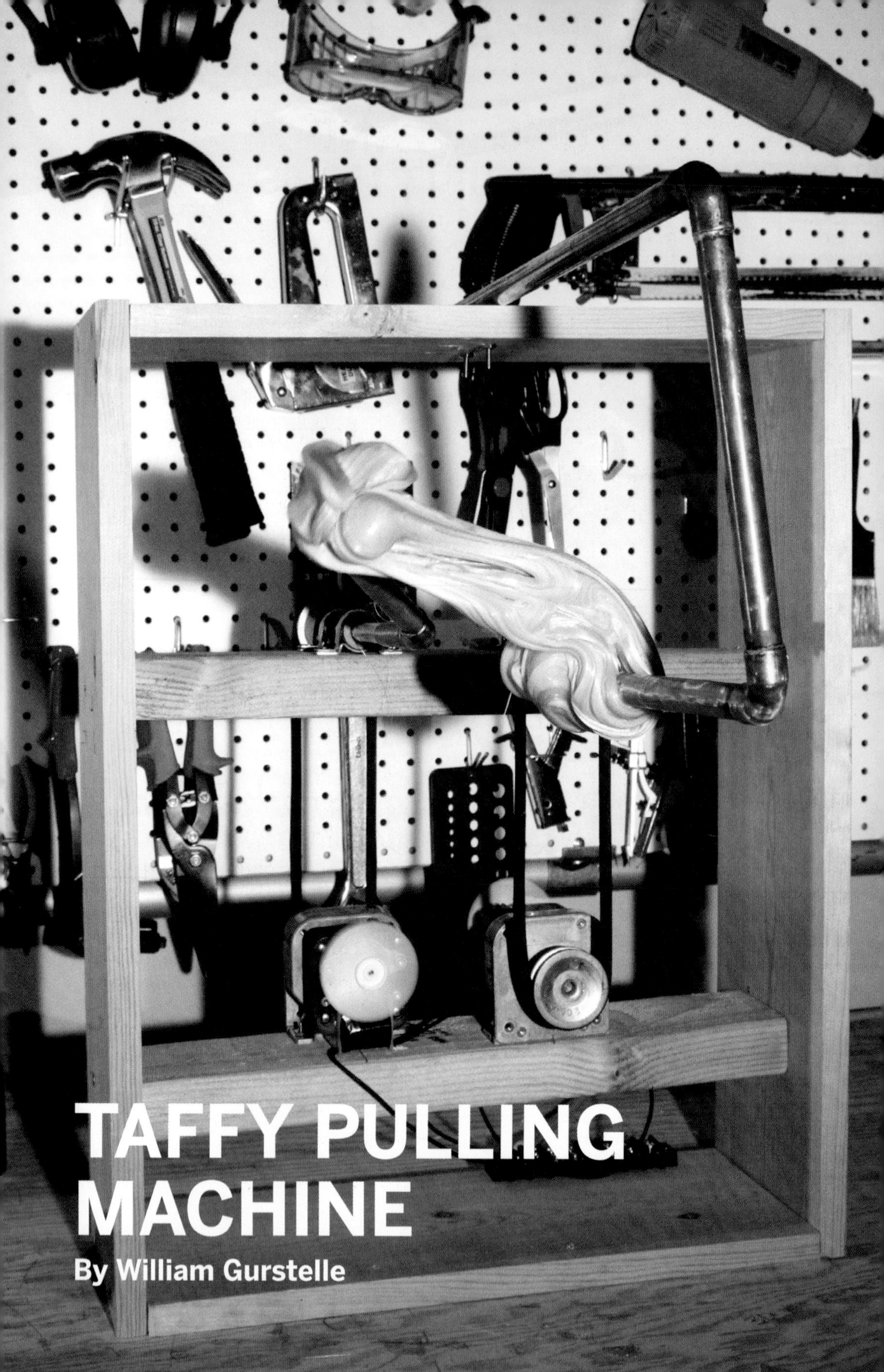

TAFFY PULLING MACHINE

By William Gurstelle

TWISTED, TUGGED, AND TASTY

Make a simple mechanism that pulls delicious candy while it stretches the limits of multidimensional math.

Sometimes the simplest things have richer histories and more complex scientific connections than you ever imagined. Such is the case for the taffy pulling machine. Taffy pullers manipulate long strands of semisolid, sugary dough into the delicious, chewy confection called salt water taffy. After viewing myriad taffy machines in tourist traps and internet videos, I sought to build one for myself.

Why? Well, when in operation, these devices display a periodic, repetitive motion so mesmerizing that, when placed in the front window of a candy shop, they invariably attract large audiences and pull in customers.

But there's more here than just a strangely compelling visual experience. I believe the appeal is due in part to a subliminal appreciation for the complex math lurking in the motion of the taffy pulling machine. In the turning arms of the engine, you'll find mathematics complicated enough to be the basis of a Ph.D. thesis. In fact, a number of academicians have investigated the hypnotic, repetitive motion of the taffy pulling machine as a portal to advanced mathematical research.

Photograph by Sam Murphy

Set up: p.110 **Make it:** p.111 **Use it:** p.113

William Gurstelle is a contributing editor for MAKE and the author of *Backyard Ballistics* and *Whoosh Boom Splat*. Visit williamgurstelle.com for more information on this and other maker-friendly projects.

A CONFECTION IN CHAOS

Dr. James Yorke is chair of the mathematics department at the University of Maryland and the man who introduced the term "chaos theory" into the popular lexicon. Yorke explains that the machine manipulates taffy in a unique form of stretching, free from tight-radius turns and pinch points.

"In each cycle of the rotors, each 360-degree turn, the taffy stretches apart by a factor of six," explains Yorke. "It becomes six times longer and six times thinner. The taffy machine carries out this stretching procedure without ever folding the taffy, without making any kinks or hairpin points. That's the feature that attracts us, because in chaos theory, there are many such situations where we abstract something that's stretched and then folded."

By modeling the stretching involved in taffy pulling, Yorke and his colleagues forged several important advances in the field of mathematical dynamics.

Understanding those advances involves discussions of "abstract chaotic dynamical processes" and "the homeomorphism of compact surfaces," so the layperson desiring more insight must await a different article written by an author with a more academic bent than me.

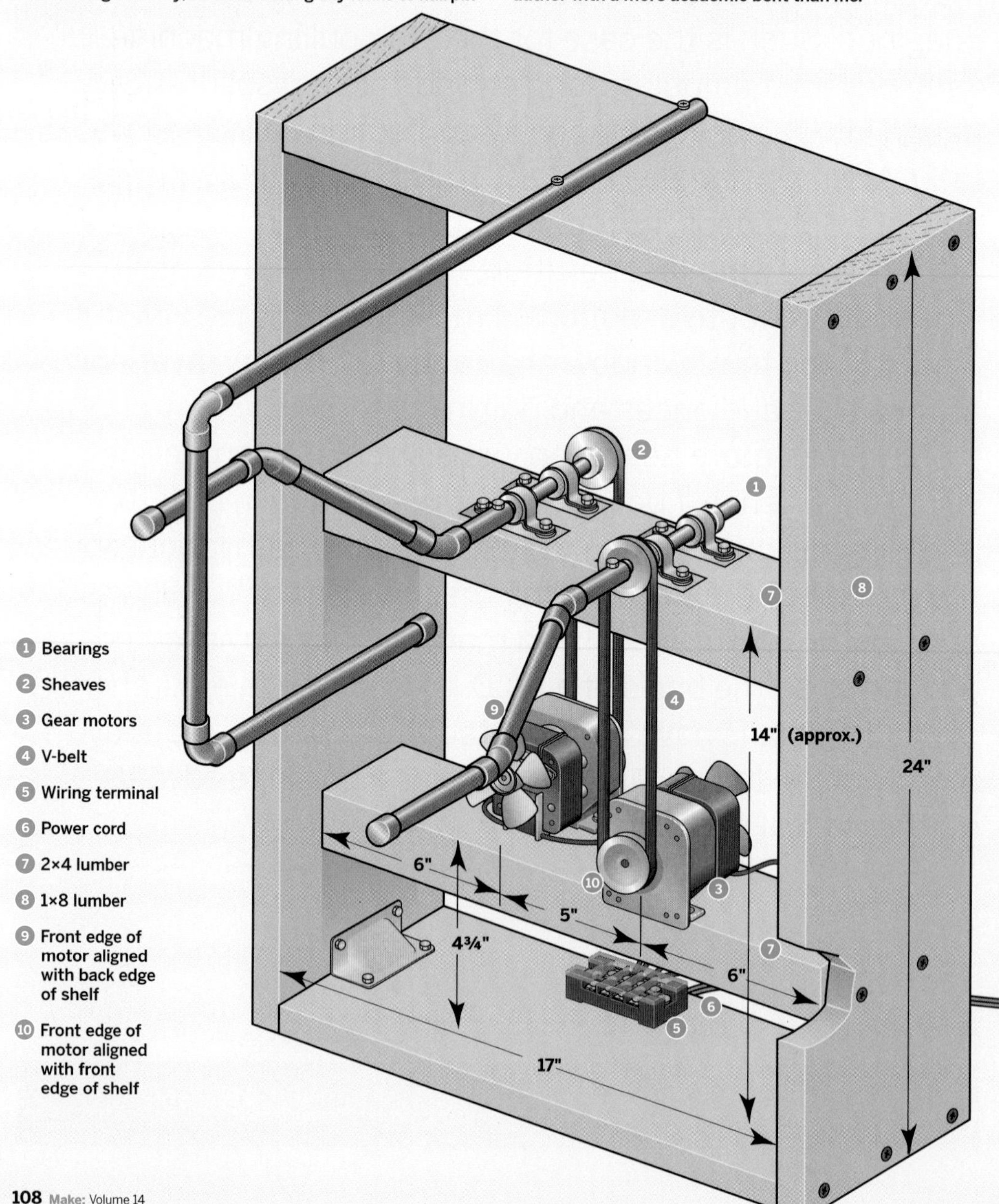

Illustration by Bill Oetinger

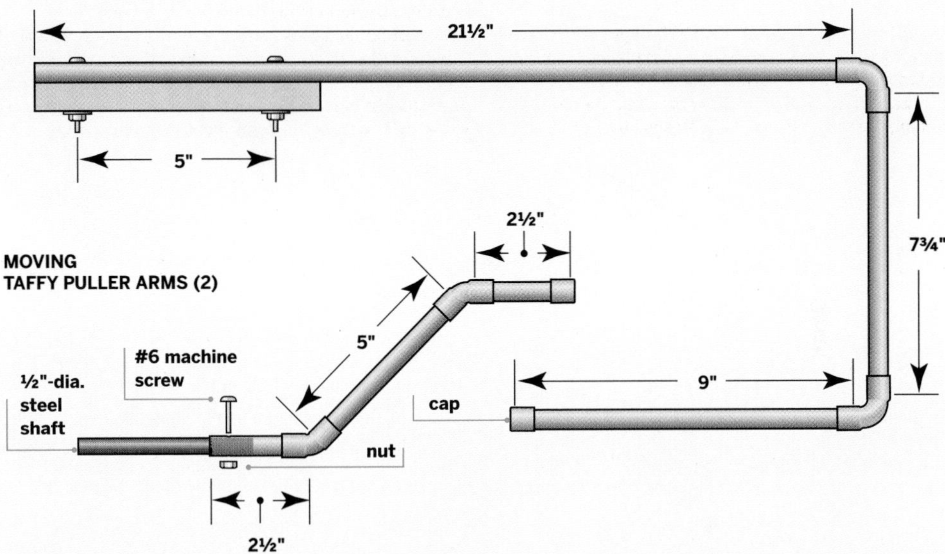

TAFFY 101

I take it on faith that taffy machines answer some important mathematical questions. But more importantly, I think taffy machines just look cool.

Taffy pulling machines appear to have emerged on the American candy scene in the late 1800s when salt water taffy rocketed to the heights of sweet, sweet popularity, based partially on its delicious, dental-filling-removing chewiness, and partially on the manner in which it was marketed. Hypnotic taffy-making machines, placed in store windows in seaside resorts, produced in the viewer a nearly involuntary urge to rush in and buy the colorful candy confection, without regard to price or nutrition.

Salt water taffy became the classic souvenir of many a trip to the beach. While taffy was sold in other places before its close association with Atlantic City, N.J., it was in the shops on that city's famous Boardwalk that it gained the "salt water" moniker and became ensconced as an American institution.

In the 1880s, Joseph Fralinger, a former glass blower, fish merchant, and baseball club manager, opened a store on the Boardwalk selling lemonade and cigars. But he soon came up with a better idea: selling multiple flavors of salt water taffy to the hungry hordes, conveniently boxed for travel and gift-giving. He had a ready market in the great crowds of tourists visiting the Jersey shore to escape the summer heat and grim urbanity of the great cities of the Eastern Seaboard. Soon this sideline turned into a full-time job. Fralinger turned his attention to perfecting his taffy recipes, first using molasses, then vanilla and chocolate. Eager vacationers lined up to enjoy the delicious culinary chaos in each chunk of his freshly pulled taffy.

Multidimensional topological mathematics and consumer marketing strategies notwithstanding, making your own taffy pulling machine is a worthy maker project. Engineering the wonderful rhythmic motion of the machine's moving parts is the primary goal of the project, and there are several ways to accomplish this. I chose to use gear motors and a belt drive to turn the taffy puller's arms, but ambitious makers may wish to modify and perhaps improve on these plans by using alternative drivetrain parts such as gears and hand cranks. An on/off switch would be one obvious improvement.

The parts I selected are not terribly expensive, and makers with a knack for scrounging may be able to find them in surplus stores or online auctions, possibly reducing the cost of the machine.

SET UP.

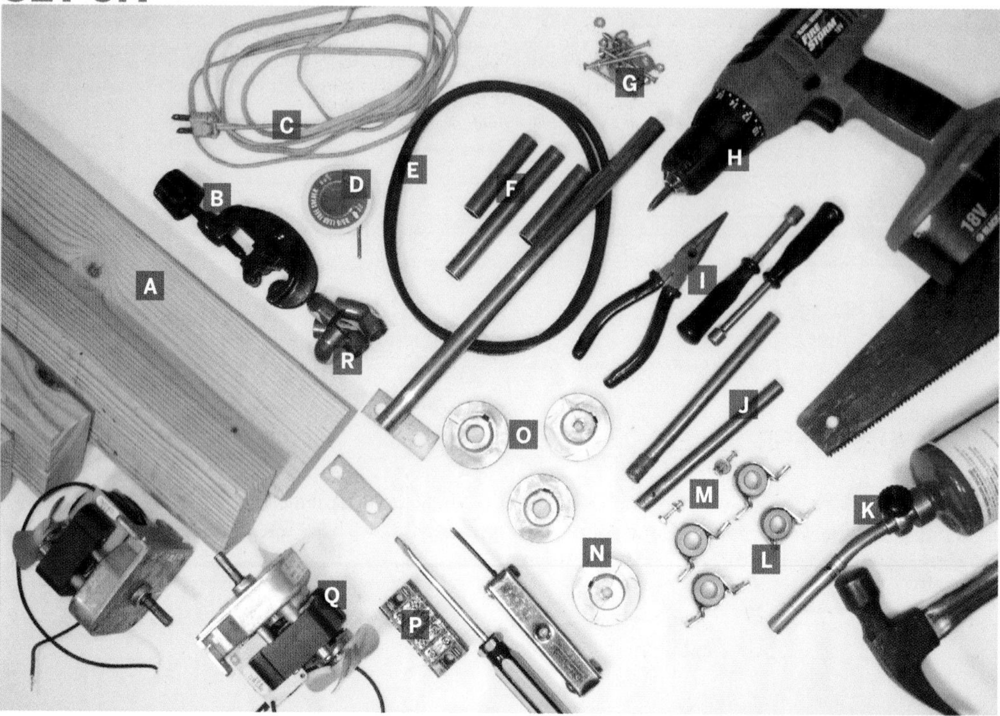

MATERIALS

[A] Lumber:
2×4 stud at least 6' long
1×8 pine board, 8' long

[B] Tubing cutter

[C] Electrical cord and plug

[D] Lead-free rosin core solder

[E] 24" V-belts (2)

[F] ½"-diameter rigid copper pipe, 8' long **for the taffy puller arms. Several types of copper pipe are sold in hardware stores: rigid and flexible, in different thicknesses designated as K, L, and M. Generally, L pipe is used for indoor and outdoor water and gas plumbing, while the thinner M pipe is used for indoor water distribution. I chose L pipe so the arms would be sturdier.**

[G] #10 machine screws with washers and nuts plus deck screws, wood screws, or nails **for the frame construction**

[H] Electric drill with drill and driver bits

[I] Standard handyperson's tools and supplies **including a saw, a hammer, pliers, screwdrivers, hex wrenches**

[J] ½"-diameter steel shafts, 9" long (2)

[K] Propane torch

[L] Base-mounted bronze pillow block bearings, ½" bore (4) **Standard bronze bearings are sufficient; ball bearings are probably overkill for this project.**

[M] 1" #6 machine screws with washers and nuts

[N] V-belt sheaves, 2" outside diameter, ½" bore (2) **These are pulleys designed to fit the V-belts listed.**

[O] V-belt sheaves, 2" outside diameter (2) with inside bore that matches the drive shaft of the gear motors. **For example, if your gear motor has a 5/16" output shaft, then the sheave bore must be 5/16".**

[P] Electrical terminal strip **insulated and covered**

[Q] 20rpm gear motors, base-mounted, 115V AC (2) **with 20 inch-pounds of torque or better. The higher the torque, the larger the wad of taffy you can pull.**

[R] Copper fittings, ½" diameter:
90° elbow (2)
45° elbow (4)
End cap fittings (3)

[NOT SHOWN]

Drilling/cutting fluid

Copper pipe soldering supplies **including emery cloth, flux, pipe cleaning brush**

NOTE: I ordered the drivetrain parts from McMaster-Carr (mcmaster.com) because they were easily available and not terribly expensive. You can also obtain parts from a local supplier, a surplus dealer, or another mail order supply firm. If you do, be sure the parts you order are compatible with one another. For example, the sheaves must have the correct profile (2L, 3L, etc.) for the belt you order, and the bore of the bearings must match the diameter of the 18" shaft.

Photography by William Gurstelle

MAKE IT.

BUILD YOUR OWN TAFFY PULLER

START Time: **A Weekend** Complexity: **Medium**

1. BUILD THE FRAME

I built the taffy puller frame out of 1×8 and 2×4 dimensional lumber. First, cut all lumber to the lengths shown in the illustration on page 108 and square them on a table saw. Then build the outer box as shown, using deck screws to make the frame solid and tight.

Next, attach the bottom 2×4 shelf to the frame. The shelves are set back from the edge of the frame to allow you to install a cover made of clear rigid plastic if desired. Don't attach the top shelf yet.

2. ASSEMBLE THE DRIVETRAIN

Assemble the drivetrain as shown in the illustration on page 108. Notice that one gear motor faces forward and the other faces backward. This allows the taffy-pulling arms to rotate in opposing directions. The placement of the bearings on the shelf is crucial. If the arms are too close together, they'll collide as they rotate. If they're too far apart, they won't fold the taffy properly.

Also, the center-to-center distance of the sheaves is critical. If the distance is too great, you won't be able to install the belt; if it's too small, the belt will sag and there won't be enough tension to keep it from slipping. You should test-fit the top shelf before mounting it, and you may have to place shims under the bearing mounts to keep enough tension on the belt to make it drive smoothly.

3. FABRICATE THE TAFFY-PULLING ARMS

This step involves soldering or "sweating" several copper fittings and pipes to form the arms that pull the taffy. If you've never soldered copper pipe before, don't worry; it's an easy-to-learn and useful skill. Excellent tutorials are available in how-to books and on the internet; typing "how to solder copper pipe" into any search engine will get you started.

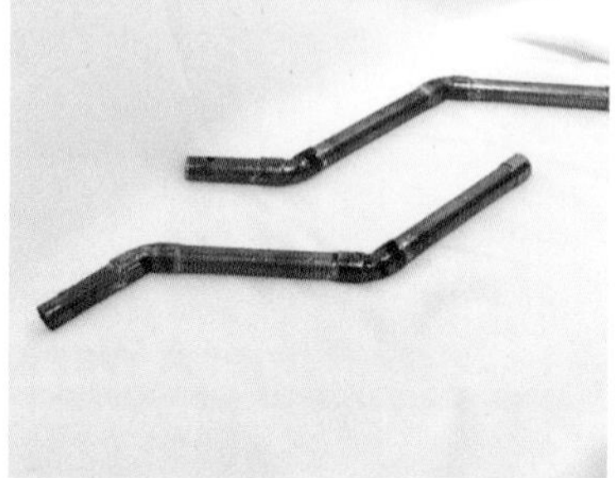

NOTE: You must use 45° elbows so that the arms can rotate without interfering with one another.

Refer to the illustration on page 109 for direction on how to solder the parts together.

Attaching the taffy-pulling arms to the ½" steel drive shaft can be somewhat tricky. I used my trusty drill press to drill a 9⁄64" hole through both the copper tube and the ½" shaft, and then inserted a 1" #6 machine screw to connect them. Take your time when drilling; this is a fairly deep hole through solid steel. Use cutting fluid, drill at a low speed, and frequently back the drill bit out to remove chips.

4. FABRICATE AND ATTACH THE STATIONARY ARM

Sweat together a copper pipe construction as shown in the illustration on page 109. When cooled, drill two or three 5⁄32" holes in the assembly and wood frame as shown in the illustration on page 108. Attach the stationary arm to the wood frame with #10 machine screws, washers, and nuts.

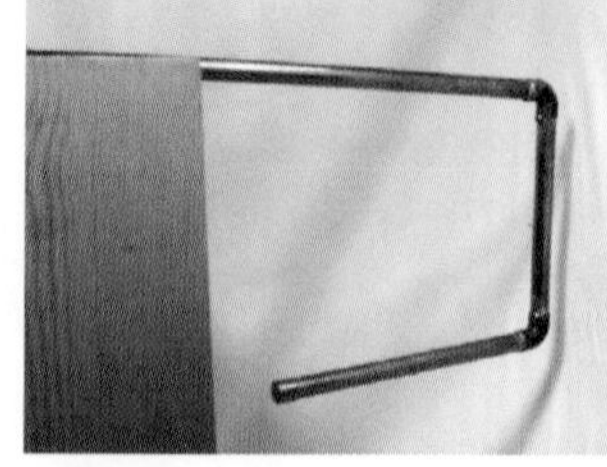

NOTE: The positioning of the stationary arm is critical. Both moving arms should revolve around the stationary arm without touching. If the moving arms do touch the stationary one, you need to realign your machine until they pass freely.

5. ATTACH THE ELECTRICAL CORD

Attach the electrical cord and the motor leads to an insulated, covered electrical terminal strip. Test the taffy puller by plugging it in.

Make any alignment adjustments necessary to insure smooth rotation of the arms. Once aligned, unplug the machine from the wall outlet. Your taffy pulling machine is now ready for its maiden stretch!

CAUTION: **120V AC current is dangerous and misuse can injure or kill. Seek the assistance of a qualified electrician if needed. Although the arms of the machine rotate at relatively slow speed, the complex motion of the arms constitutes a pinching hazard. Keep fingers well away from the pinch points.**

FINISH X

NOW GO USE IT

USE IT.

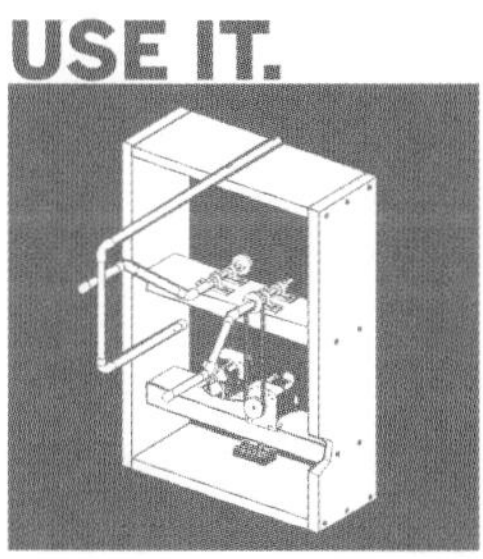

OPERATING YOUR PULLER

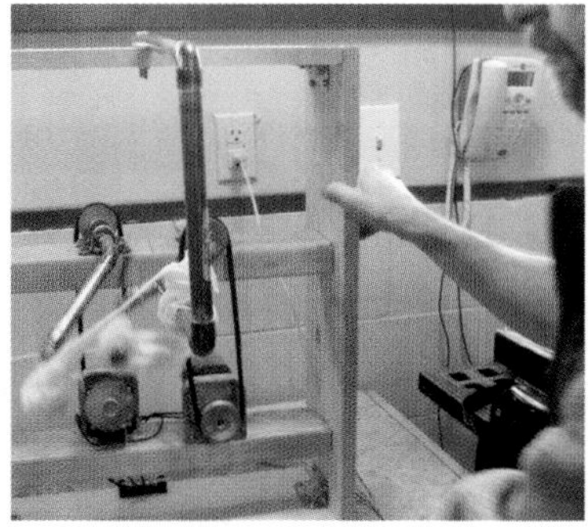

RECIPES AND TIPS

There are many recipes for taffy on the internet. You may need to try a few before you get one that works well with your machine.

Once the taffy is cooked, I pull it by hand to get it started. I plug in the puller, transfer a lump of taffy weighing roughly 5oz to the machine, adjust it until it stays on the arms, and let the machine do its job. Be careful not to get pinched by the moving machinery!

The key to good taffy is timing. The ingredients must be watched closely as they heat to a precise temperature known in candy making as the "hard ball stage." The taffy mixture is allowed to cool until it can just be handled and then is placed on the machine and pulled. Working too quickly or too slowly will make your taffy stringy, lumpy, or gluey.

The taffy becomes thicker as it's folded in the machine. Toward the end of your taffy pull, you may have to give the rotating arms an occasional shove to keep the taffy moving. Keep experimenting. Whether it's the mesmerizing repetitive motion of the machine or the prospect of a sweet treat, soon you too will be pulling in an appreciative crowd.

THE PIXELMUSIC 3000

By Tarikh Korula

TV PARTY TONIGHT

Re-create a mid-1970s video trip by plugging this box into any TV and audio source. Beneath the fake wood paneling, a Propeller microcontroller simulates Atari's classic music visualizer.

In 1976, Atari introduced Atari Video Music, a plugged-in music visualizer designed by Pong creator Bob Brown that bridged the yawning gap between consumers' stereos and their TV sets. The quirky, psychedelic pixelation device never caught on, but watching it in action today (search YouTube), one is taken back to another time, long before iTunes and Winamp visualizers. It was a time when vinyl, denim, Foghat, mood rings, limited color palettes, and RadioShack's business model all somehow made sense.

And while Foghat's career may be a distant memory, interest in Atari's long-gone device remains. So we introduce the Pixelmusic 3000, a weekend project that pays tribute to those groovy times, and to a product that was either too quirky or too revolutionary to make it past its first year's production run.

Today, of course, the technologies that enabled Atari Video Music are much smaller, cheaper, and more accessible. We'll use the Propeller microcontroller and its video libraries to create a simple AVM-like visualizer that feeds a TV from an iPod or other music player.

Photography by Tarikh Korula

Set up: p.117 Make it: p.118 Use it: p.123

Tarikh Korula is co-founder of the software and hardware development firm Uncommon Projects (uncommonprojects.com).

BEHIND THE MUSIC VISUALIZER

The $12 Propeller microcontroller lacks sufficient RAM for a frame buffer, so it can't produce any-and-all video imagery. But it's powerful enough to synthesize a real-time NTSC television signal that renders geometric shapes and colors.

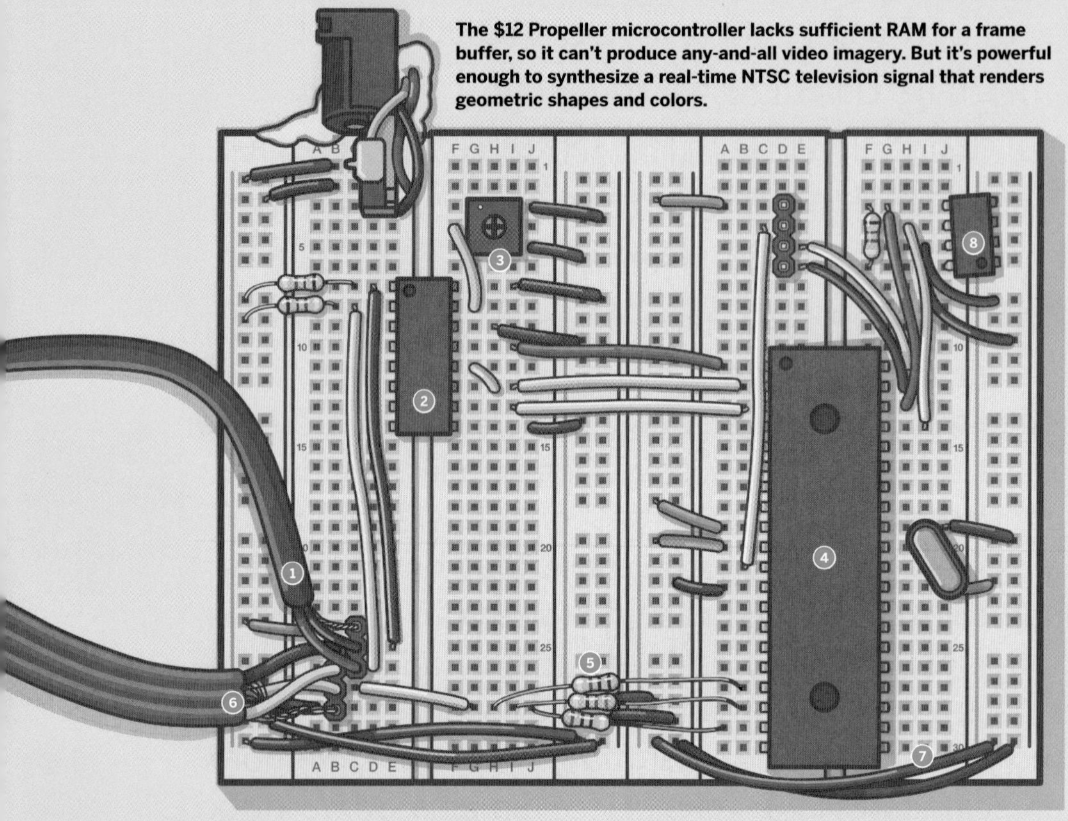

HARDWARE

1 Stereo signal from the audio cable is split on the circuit board. Half is carried back out by the A/V cable, and the other half feeds the analog-to-digital converter (ADC).

2 The ADC chip converts analog value levels into digital values that are usable by the microprocessor.

3 The potentiometer lets you manually adjust the reference voltage that the ADC compares left and right channel volume levels to. You may find small increments at the edge have a big effect.

4 The Propeller microprocessor runs software (see below) to create visuals based on input from the ADC (pins P0–P2). It outputs them (pins P12–P14) as a digital precursor to NTSC video.

5 The resistor ladder divides and recombines 3 digital outputs from the microprocessor to produce an analog NTSC video signal, the pre-HD standard for TVs in North America, Japan, and elsewhere.

6 The A/V cable takes the original stereo signal and the generated NTSC video out to a TV or home entertainment system.

7 Pin 16 of the microprocessor (not shown) connects to a red indicator LED (not shown on breadboard). The software switches the LED on to show that it's running successfully.

8 The EEPROM chip (electrically erasable programmable read-only memory) stores the software on the board between power-ups. The chip is intentionally installed upside down on the board so you can easily ground 4 of its pins at once.

SOFTWARE

The Pixelmusic software "paints" simple shape layers using a limited palette of 70s-compatible colors. Their size, orientation, and color palette are controlled by volume changes in the left and right channels of the music. One of 8 different shape layouts is randomly selected each time there's a sudden change in volume. Like the original AVM, the Pixelmusic doesn't look at frequencies, beats, or anything fancy; just the left and right channel volumes. But because our brains are wired to make audio-visual correlations, this is enough to make people swear that the patterns are dancing to the music.

Illustration by Damien Scogin

SET UP.

MATERIALS

[A] Stereo mini (3.5mm) cable

[B] Mini-to-RCA A/V cable

[C] Propeller 40-pin microcontroller **Parallax #P8X32A-D40**

[D] 24LC256 256K serial EEPROM memory

[E] Microchip MCP3208 analog-to-digital converter (ADC)

[F] LM2937 3.3V voltage regulator

[G] 5MHz crystal

[H] Capacitors: 0.1µF and 22µF

[I] Resistors: 270Ω (2), 560Ω, 1.1kΩ, 4.7kΩ (2), and 10kΩ

[J] Red LED

[K] Circuit board headers: 3-pin and 4-pin

[L] 40-pin IC socket

[M] Perf board **RadioShack #276-150**

[N] 6V 300mA DC power supply **"wall wart"**

[O] DC power jack to PCB adapter **that fits wall-wart plug, Digi-Key #CP-202A-ND**

[P] Ribbon cable **any width**

[Q] 22-gauge solid hookup wire **various colors**

[R] Prop Plug programming connector **Parallax #32201**

[NOT SHOWN]

Slide switch

10kΩ trimpot variable resistor **aka potentiometer**

Serpac A-21 enclosure, black **Jameco #373333**

Wood-grain contact paper **from a local hardware store**

Rubber feet (4)

TOOLS

[S] Wire cutters

[T] Wire strippers

[U] Needlenose pliers

[V] Multimeter

[W] Hobby knife

[X] Precision screwdriver set

[Y] Soldering iron and solder

[Z] Hot glue and glue gun

[a] Electrical tape

[b] Helping hands

[c] Cutting mat/ work surface

[NOT SHOWN]

Solderless breadboard **RadioShack #276-003; I used 2 modules.**

Solder wick or desoldering tool

Dremel tool with cutting, grinding, or routing bits

Windows XP/Vista computer

MAKE IT.

BUILD YOUR PIXELMUSIC 3000

START **Time: A Weekend Complexity: Medium**

1. MAKE THE CABLES

If you've never hacked a cable before, there's a rush when you make that irrevocable snip.

1a. Cut the stereo mini cable near 1 end, and strip the wires 2" down. Use a multimeter's continuity function to identify left channel, right channel, and ground (GND) wires, which correspond to the jack's tip, ring, and sleeve, respectively. Ground may be a mesh surrounding the other 2 wires.

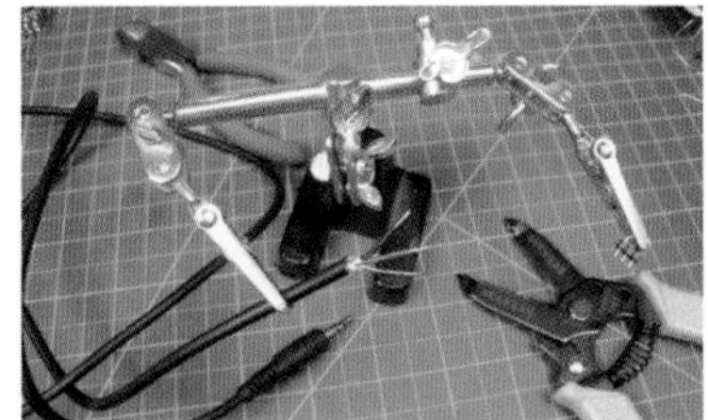

1b. Cut the mini-to-RCA A/V cable near the mini end, then strip and identify the wires as you did with the stereo cable. This cable has an extra ring for video, which connects to the yellow RCA plug.

1c. Solder the cable wires to the 3- and 4-pin circuit board headers. With the stereo cable, solder in this order: ground, right channel, left channel. With the A/V cable, follow the order ground, video, left, right. Confirm all connections with the multimeter — inside my possibly nonstandard A/V cable from a 99¢ store, the white and yellow wires were reversed.

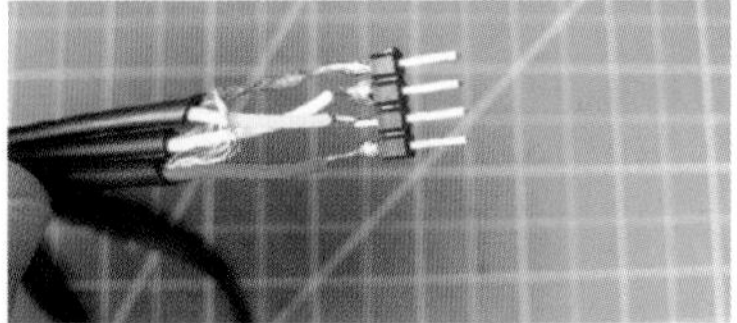

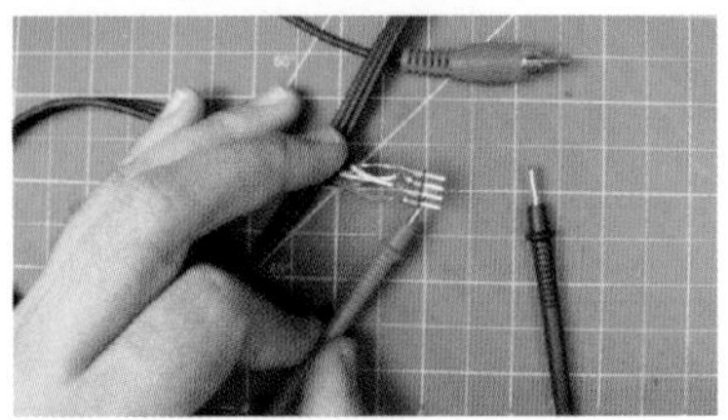

2. BUILD THE BREADBOARD AND TEST THE CIRCUIT

2a. Insert the Propeller microcontroller, ADC, and EEPROM chips into your breadboard. I used 2 breadboard modules, with the Propeller over one centerline and the ADC over the other. As a rule, the top ends of the chips, indicated by a notch, should point toward the top of the board, like north on a map. Pins are numbered counterclockwise from top left. (However, I inserted the EEPROM chip upside down so I could ground its pins 1–4 in one move, using one of the breadboard's bus strips.)

TIP: Chips ship with their pins slightly splayed, so before you put them into a breadboard, rock them on a hard, flat surface to turn the pins in slightly.

2b. Follow the schematic at makezine.com/14/pixelmusic to assemble the rest of the circuit on your breadboard using hookup wire. This lets you debug the circuit before committing it to solder. Other components include your 2 cables, a voltage regulator, a trimpot, a crystal, and a header for the Prop Plug programming port. I hot-glued the DC power adapter to the edge of the breadboard near the voltage regulator. Our LM2937 regulator has different pin assignments from the standard 7805 you may be used to, so check the datasheet on yours.

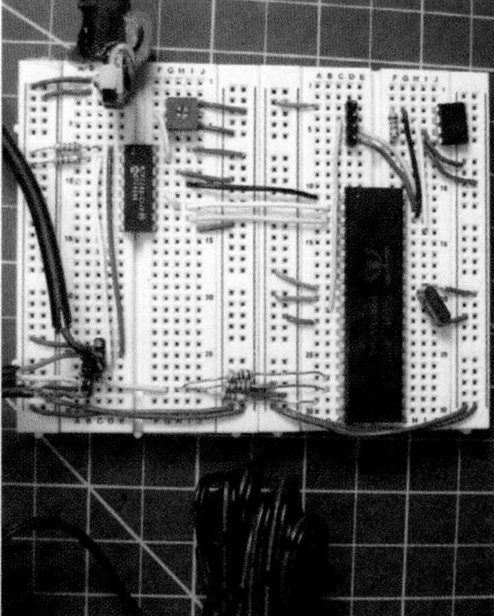

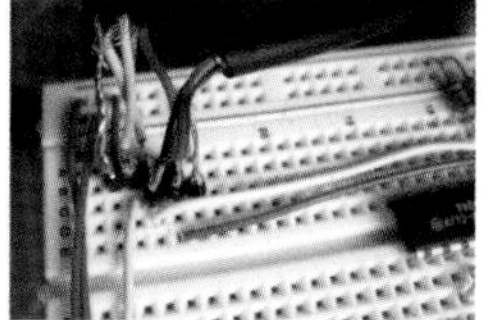

TIP: Both here and on the soldered circuit board, with few exceptions, I use blue wires for ground, red for power, and white, yellow, or black for other connections. Following a system like this makes the wiring easier and helps you visualize designing your own 2-layer PC boards.

2c. If you haven't done so already, install the Propeller Tool IDE (integrated development environment) on your PC. It's a free download from parallax.com.

2d. Run the Prop Plug from your PC's USB port to the circuit, connect the DC power to the breadboard, and launch Propeller Tool. Hit F7 to confirm that it sees a Propeller chip. If not, check all connections, unplug and replug the Prop Plug, and try again. Standard debugging rules apply.

2e. Download our *pixelmusic.spin* program file from makezine.com/14/pixelmusic, open the program in Propeller Tool, then upload it onto your Propeller by pressing F10. Presto! You can stop now, plug this thing into your TV and iPod, and have a personal dance party! The rest of this tutorial is just sweet icing on the cake.

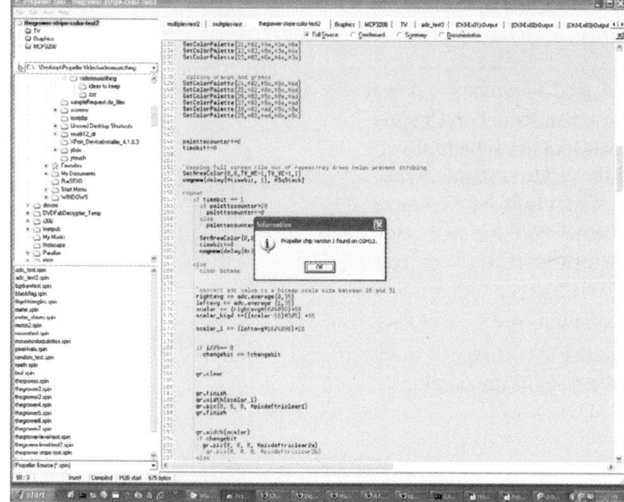

3. SOLDER THE CIRCUIT BOARD

If you're new to soldering, proceed with caution; there's a lot of minute work here. Also, there's no shame in using a bigger perf board — it will make your life easier.

3a. Add the 40-pin socket to the board, over the centerline, and solder 2 corners down. You won't solder every pin; on such a tight board, you only want to solder the pins being used.

3b. Add the EEPROM and crystal. Orient the EEPROM upside down, with the pins nearest the notch (5 and 6) facing the bottom of the board and connecting to pins 37 and 38 of the socket. Wire pin 8 to the middle rail for power, and solder-bridge pins 1–4 together and connect them to the rail for ground. Connect the crystal to pins 30 and 31 of the socket.

NOTE: For each connection, fill both holes with solder, then span them with a wire. Or for short straight runs with nothing in between, you can use a solder bridge. Test the continuity to confirm connections and check for shorts.

3c. Add the 10kΩ resistor. Wire one side to EEPROM pin 5 and Propeller pin 38. Wire the other side to power and bridge the connection over to the EEPROM's power pin, pin 8 (under the blue ground wire).

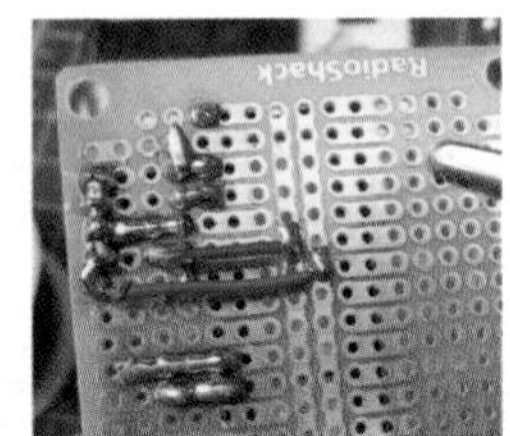

3d. Add all remaining power and ground connections for the Propeller, following the schematic.

3e. Solder the 4-pin programming header to the edge of the board and wire all of its connections.

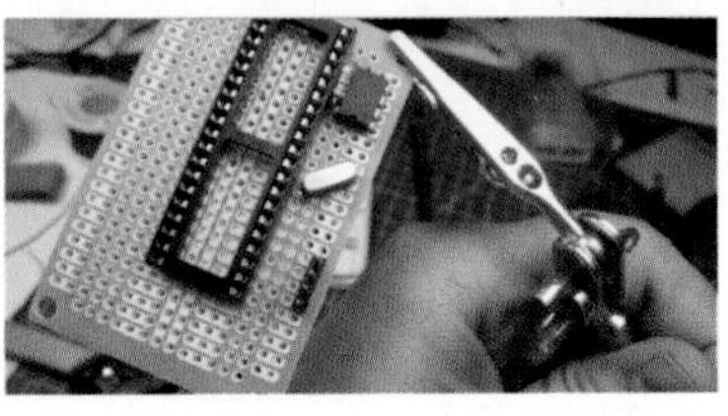

3f. Solder leads from the DC power adapter to the board, connect them to the voltage regulator, and add a 1µF decoupling capacitor between power and ground. I used a smaller nonpolarized cap to save space, but if you use an electrolytic, follow its polarity. Plug the adapter in and use your multimeter to confirm that you have identified power and ground correctly; reversing these will fry your chips.

3g. You should now be able to program your Propeller through the perf board. Insert it into the 40-pin socket, making sure all pins seat correctly in the holes. Use a thumb on each side to press the chip down into place. Connect the Prop Plug to the board's programming header, and connect the DC power plug to the adapter jack.

NOTE: Run Propeller Tool on your computer and confirm that it can see the chip by pressing F7. Then reprogram the chip by pressing F10, and finally, upload the code to the EEPROM by pressing F11.

3h. On the upper left of the board, solder and wire up the ADC chip and the trimpot, following the schematic.

3i. On the lower left, follow the schematic to create and connect the resistor ladder (270Ω, 550Ω, and 1.1kΩ resistors) and the stereo and A/V cable headers. I used red and white wires on the topside of the board to send the stereo signals up to pins 1 and 2 of the ADC.

3j. Use electrical tape and hot glue to provide insulation and strain relief on the cables, which you'll bend pretty hard to fit into the case.

3k. Connect a red LED to the circuit via a 2-conductor strip of ribbon cable. Wire the + side of the LED (the long leg) to Propeller pin 21 through a 270Ω resistor, and the – side to ground. The flexible cable lets you position the LED inside the case.

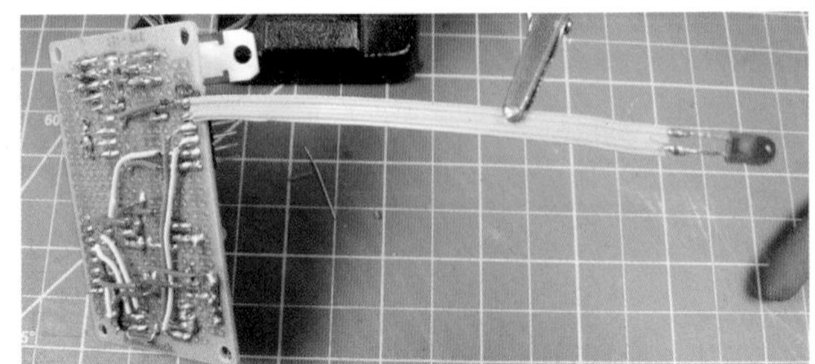

4. BUILD THE BOX

Nothing evokes 70s product design like injection-molded plastic, fake wood paneling, and a red LED.

4a. Measure the Serpac case for holes for the cables, LED, and power switch. Then carefully carve them out with a Dremel. I put the switch in one of the side panels. Then trace the case's side panels on the contact paper, cut with a hobby knife, peel, and stick. Smooth!

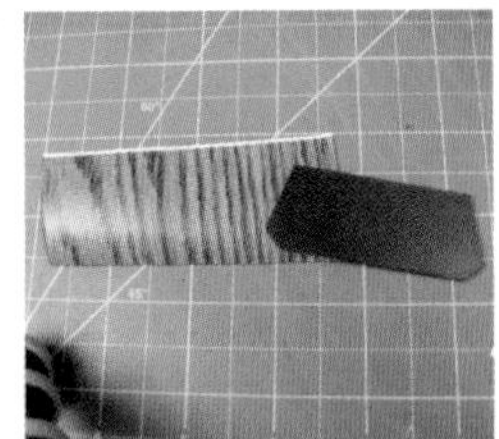

4b. Hot-glue the slide switch into the case and splice its contacts into the + power lead. Use more hot glue to fix the LED and cables in their holes and stick the circuit board to the case bottom. I contend that hot glue is integral to the success of this (any?) hacking project. It's nonconductive, fast, and you can remove it easily if you make a mistake.

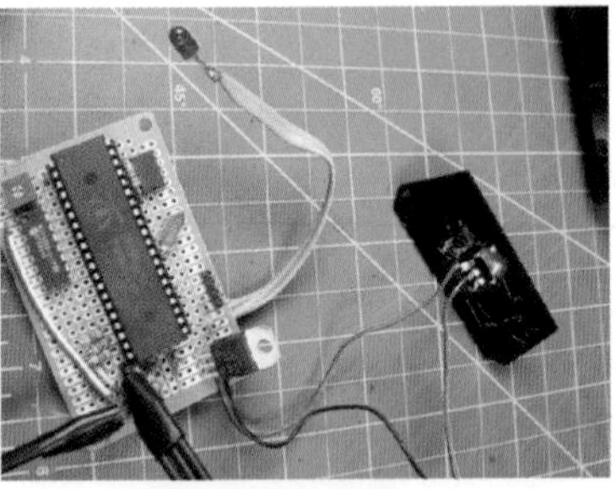

4c. After confirming that the circuit works, screw the case closed, add rubber feet, and put on any finishing fake wood touches. Attention to detail is proportional to the amount of marvel in your friends' eyes the first time they see your Pixelmusic 3000 in action.

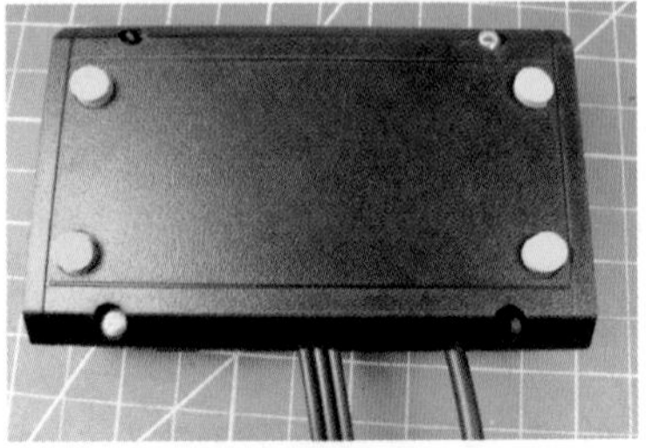

FINISH X

NOW GO USE IT »

USE IT.

HOOK UP AND ROCK OUT!

THE TELEVISION CONNECTION

With any luck, you're now the proud parent of your very own Pixelmusic 3000. Perhaps you stopped with the Breadboard 3000, or maybe you went all the way to enclosed-perf-board heaven. They both work.

The PM3K sends a composite NTSC signal to the TV over the yellow RCA cable. Older sets, like the beauty pictured below, may not have a composite in (ours didn't even have a coaxial in). Not to fear; just head down to any electronics store and get an RF modulator, which will convert the composite signal to 2-wire VHF antenna input on channel 3 or 4.

HOW TO USE

There's a sweet spot for visualizations. Too little volume and the shapes, patterns, and colors won't change very much. Too much volume and things may change too frequently or be too big. Try playing with the input volume to find just the right level. You may also want to adjust the ADC's trimpot to make big tweaks once or twice.

MODS AND ROCKERS

The goal of this project is to create a retro rave in your living room, but the PM3K is portable too, so you can bring it to your friends' parties and impress all the right people. It works with iPods, CD players, or anything else with a mini headphone output jack, and it's just as fun with modern music and audio books as it is with classic album-oriented rock.

There's plenty of room to mod your PM3K. The code is yours to tweak and revise, and it's fun trying to create programs to run with the limited RAM and color budget that the Propeller allots. You might want to try adding some pots and switches; there are open slots on the ADC and open pins on the Prop. Or maybe an onboard mic? Whatever you do, be in touch and let us know. We'll drive our Nova over and bring our Freedom Rock playlist.

RESOURCES

Watch a video of the Pixelmusic 3000 in action, and tell us about your Pixelmusic build, at makezine.com/14/pixelmusic.

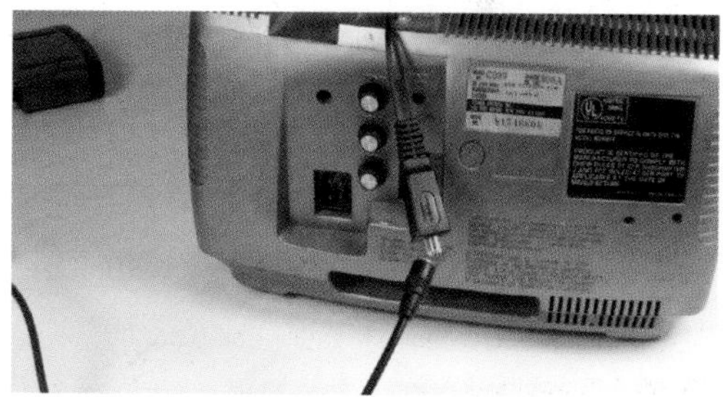

MAKE MONEY

Piggy Bank

By Tom Parker

Sometimes it costs more to buy it than to make it from the money itself.

$43.00

Pottery piggy bank on eBay.

↑ $25.01

Piggy bank made entirely out of money (25 bills with one penny added for the tail).

DIY CIRCUITS

WIRELESS MOTION SENSING MADE EASY

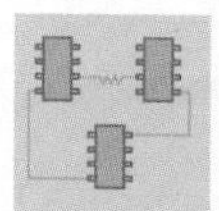

XBee radios track every hit in roller derby action! By Tom Igoe

I was never much of a sports fan until my friend Mattie introduced me to the Gotham Girls Roller Derby (gothamgirlsrollerderby.com). Once you've seen a good blocker send the opposing team's jammer sailing into the sidelines with a solid hip check, you're hooked. Add all the bad puns in the players' nicknames, and you've got a sport I just can't resist.

Soon, I had to find a way to hack it. If the skaters wore motion sensors, I figured, you could make things react to the action. Sound effects on every hit! A synchronized soundtrack! Flames shooting up every time a player gets knocked out of bounds! The possibilities are endless.

At first I thought of giving each skater a Wii Remote, but the signal isn't strong enough for an 88'-long track. I still wanted to keep it simple, so I paired a 3-axis accelerometer module with an XBee radio from Digi. The basic XBee is great for wireless sensing projects; configuration is easy and its onboard inputs mean you don't need a microcontroller. So, the final part count for each skater module is 3 components and a couple of sockets on a small board.

In addition to the skater radios, you'll need a base radio attached to a computer. I outfitted 3 skaters with radios and set up my laptop to graph the data and play sound effects when the players collided.

Photography by Tom Igoe

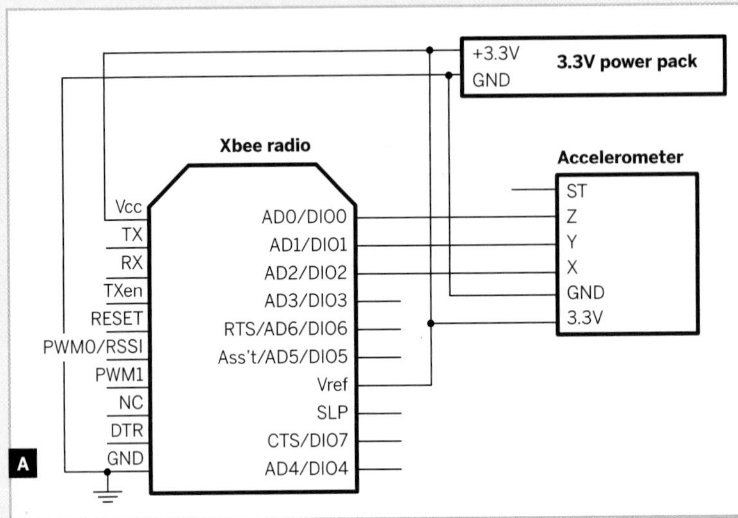

Fig. A: Module wiring: The accelerometer's X, Y, and Z outputs connect to the XBee's analog 0, 1, and 2 inputs.

MATERIALS

For each skater module:

Digi XBee 802.15.4 (formerly Series 1) radio part #XB24-ACI-001 from digi.com, $19

3.3V regulated power module #VPack3.3V_AA_1 from Bodhi Labs, bodhilabs.com, $11

3-axis accelerometer module #SEN-00849 from SparkFun, sparkfun.com, $30

Small perf board that you can snap in half to mount 2 radios, such as RadioShack #276-148

Breakout board for XBee SparkFun #BOB-08276

2mm 10-pin sockets (2) SparkFun #PRT-08272

Breakaway female headers 5 pins for each module, SparkFun #PRT-00115

Breakaway male headers 2 pins for each module, SparkFun #PRT-00116

Scrap wood or enclosure You need something strong to mount the circuit boards and batteries onto that the skaters can wear. I bolted the boards to scrap wood with #4-40 screws. You could also get fancy and build a nice casing.

For the base station:

Computer

XBee-to-USB module Droids part #990.002 from droids.it, $33, or New Micros #USB-XBEE-DONGLE-CARRIER from newmicros.com, $39. You can also use a general-purpose USB-to-TTL serial adapter like SparkFun #BOB-00718, $15, and wire it to the XBee on a breadboard.

For the skaters:

Contact your local branch through the Women's Flat Track Derby Association (wftda.org) and have plenty of beer on hand.

1. Build the skater modules.

The module circuit is simple (Figure A). The accelerometer's outputs connect to 3 of the XBee's inputs, and both are powered by a 3.3V battery power module. Tilting the accelerometer changes the voltage, and sudden changes, like from a fall or a hit, produce a sharp spike.

Plan your layout before you solder, and bend connected wires and leads toward each other underneath the board, to make soldering easy. With the XBee breakout board, I soldered only the corners and the pins I used, for ease of removal (Figures B–E).

2. Configure the radios.

All the USB adapters listed at left use FTDI's USB-to-serial chip, so you'll need to install the drivers from ftdichip.com. An XBee adapter lets you just plug the radio into your computer. With a generic USB-to-serial adapter, connect your computer's transmit pin (TX) to the adapter's receive (RX) pin, and its RX pin to the adapter's TX pin. Connect the adapter's 3.3V output to the radio's voltage in, and connect both of their grounds together (Figure F).

With the radio and adapter connected, run a serial terminal program on your computer to check for the new port. In Mac OS X you can use ZTerm (homepage.mac.com/dalverson/zterm), and with Windows use HyperTerminal (under Start → Programs → Accessories → Communications). The port should show up with a name like `/dev/tty.usbserial-A5001rNq` in OS X, and `COM4` or a higher number in Windows.

You may also need to upgrade the firmware on your radios. For instructions, search for "firmware upgrade example" on Rob Faludi's excellent XBee blog at faludi.com/category/xbee.

With your terminal program, configure and open the port for your USB adapter. In OS X, open ZTerm's Settings → Connection menu item and configure your connection as `9600-8-None-1` with no flow

Fig. B: The 2-pin male header at the top takes power, the 5-pin female header at right connects to the accelerometer, and black dots show the XBee breakout board position.

Fig. C: Bottom of the board, showing wires and leads bent before soldering. Fig. D: Bottom of the board, soldered. Fig. E: The finished board mounted on a piece of scrap wood to keep it all together.

control. In Windows, select File → New Connection in HyperTerminal and follow the prompts to give your connection a clever name like "FTDI," choose the new COM port, and specify `9600-8-None-1`. Finally, open the connection using the toolbar's Call button.

In your now-active terminal window, type: +++

Don't hit return (symbolized as `\r` here), just wait. In a second or so, the XBee radio should respond with:

```
OK\r
```

This puts the radio in *command mode*, which lets you configure it, but after 10 seconds idle it will revert to *data mode*, which just sends it whatever you type. So if you wait more than 10 seconds after any command, you'll need to send the +++ string again.

All the properties of the radio are configured by sending an `AT` command: `AT` followed by the parameter, followed by any modifiers, then you hit return. For example, to set the radio's address to 1, type:

```
ATMY 1\r
```

The radio responds `OK` when you send a good command, and `ERROR` when you don't. You can configure the rest of the radio's properties by entering the command string:

```
ATRE, IDAAAA, MY1, DL0, D02, D12, D22, IR50,
IT1, BD7, WR\r
```

This sets the network ID to AAAA; the radio's own address to 1; the destination (base station) address to 0; the active input pins to D0, D1, and D2; the analog sampling rate to 80 milliseconds (50 in hex), the samples per packet to 1, and a baud rate of 115,200 bits per second (7 is typical code). The `RE` at the beginning resets the radio to defaults, and the `WR` at the end tells the radio to save these settings even when it's not powered.

You'll configure additional mobile radios following the same procedure, but use `MY2`, `MY3`, etc. for their addresses. I also found that it helped to assign them all slightly different sample rates (`IR`) between 80 and 100 milliseconds. This reduced interference when radios were close to each other, like when skaters were bunched up in a pack.

These settings change the radio's baud rate to 115,200 bps, so you'll need to reset your serial terminal setting for 115,200 bps as well after you configure the radios. After I change each radio's settings, I like to confirm it with a string like this, which echoes its current settings:

```
ATID, MY, DL, D0, D1, D2, IR, IT, BD\r
```

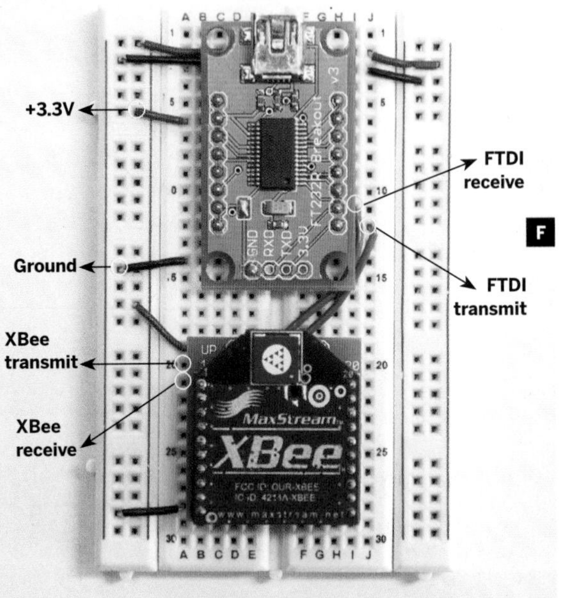

Fig. F: XBee module on the breakout board, wired to a general-purpose USB-to-TTL serial adapter. Fig. G: The Simple Processing application graphs accelerometer readings and counts a collision when they suddenly change. Fig. H: Gotham Girls fresh-meat skaters Dainty Inferno, Miss American Thighs, and Dinah Party with motion sensors. Special thanks go to them for volunteering as test subjects for this project.

After configuring all the mobile radios, use the following command to configure the base station radio to talk on network AAAA as radio 0:

```
ATRE, IDAAAA, MY0, BD7, WR\r
```

Then, using your serial terminal program, you'll need to change the port's data rate from 9,600 to 115,200, to match the radios' new rate.

3. Program and test the system.

Using Rob Faludi and Dan Shiffman's XBee API library for Processing (available from shiffman.net), I wrote a program that reads the values from multiple sensor modules, then graphs them and periodically logs them to a comma-delimited file. It also makes sounds whenever there's a collision and keeps track of how many have occurred (Figure G). See makezine.com/14/diycircuits_roller for my code.

4. Test the system *in vivo.*

You have to position the module where it won't hurt the skater if she lands on it. We tried the top of the helmet, the hip, the ankle, the shoulder, and the cleavage. The shoulder was difficult to secure to, and the radio signal got lost in the cleavage. The helmet turned out to be the safest, but the hip and ankle gave the widest range of readings and the most pronounced hits.

Padding makes the modules safer, but too much dampens the sensitivity of the accelerometer. Bubble wrap worked well for tests, but ¼" foam is more effective.

What Next?

Fun though it is, roller derby is not the only easy application for these XBee radios. You can attach any sensor that outputs voltages up to 3.3V, and control any device with the same input range. They can exchange data from computers and microcontrollers as well. I've used them to read remote photovoltaic arrays, make toxic gas sensors, and control mechanical chimps — and my students and colleagues at NYU have gone much further with them.

Thanks go to Rob Faludi, Dan Shiffman, and Kate Hartman for technical collaboration on this project, and to the Gotham Girls for their cooperation.

Tom Igoe is an associate arts professor at the Interactive Telecommunications Program at New York University. He teaches about physical interaction design, and wants to work with monkeys someday.

DIY CIRCUITS

MINI BIKE LIGHT

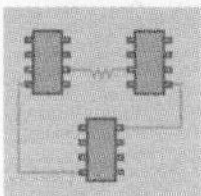

Make an easy LED headlight from a garden hose adapter. By Trevor Shannon

Photography by Trevor Shannon and Katie Dektar

I wanted to build a small, bright, and durable LED light for my bike, and I read online that plumbing parts work well as housings. So I made a 3-LED headlamp that's enclosed by a ¾" hose faucet adapter and powered by an outboard battery pack.

1. Make the LED mount.

To hold the LEDs, I used some scrap translucent plastic. Using a drill press, I marked an outline on the plastic sheet by cutting partway through with a 1¼" hole saw. I drilled three 5mm holes inside it for the LEDs, then finished cutting out the disk with the hole saw.

Trim the edges of the disk until it fits snugly in the rubber side of the hose adapter, stretching the rubber a bit. I used a grinder, but you can also use a file or sandpaper (Figure A, following page).

MATERIALS AND TOOLS

- White 5mm LEDs, 3.6V, 20mA (3) part #276-320 at RadioShack, or cheaper from mouser.com or other online suppliers
- 15Ω resistor I determined the resistor's value using a formula described in instructables.com/id/LEDs-for-Beginners.
- 3-AAA battery holder (1) with batteries (3)
- Small SPST toggle switch I used a micro-mini toggle, RadioShack #275-624, for a second, more streamlined version of the light.
- Insulated wire, electrical tape, and super glue
- ¾" garden hose faucet adapter clamp style
- ¼" hard translucent plastic small sheet or scrap
- 1¼" hose clamp and 1 extra clamp
- Soldering materials
- Drill press with 5mm bit and 1¼" hole saw
- Grinder, file, or sandpaper

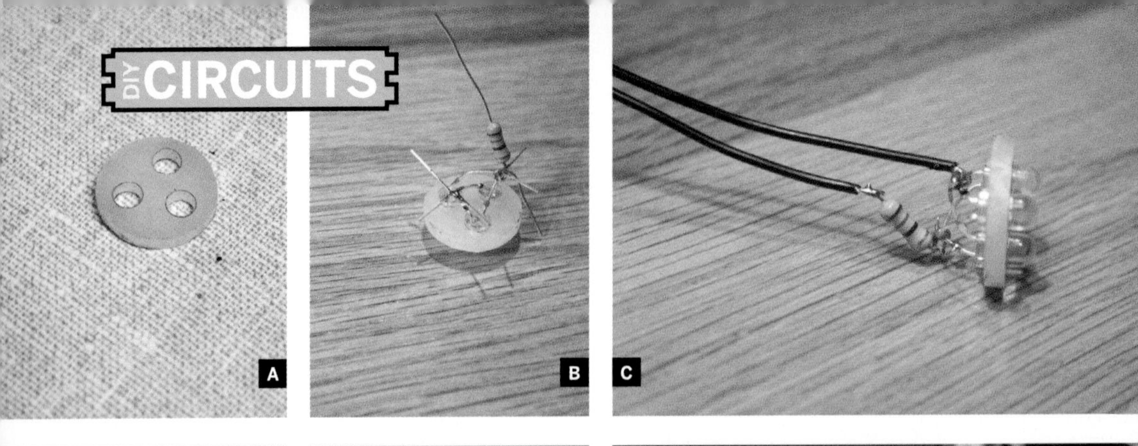

Fig. A: Drilled plastic disk for holding LEDs. Fig. B: LEDs glued into the disk and wired together with the resistor. Fig. C: Leads connected for power and switch. Fig D: LED disk fit into the rubber end of the faucet hose adapter.

Fig. E: Switch connected to the negative LED lead and negative battery terminal. Fig. F: Rider's view of the headlamp hose-clamped above the bicycle handlebar and the switch zip-tied underneath.

2. Connect the LEDs.

Super-glue the LEDs in the disk's holes, arranged so that all their short (negative) leads point toward the center. Bend and solder together the short leads, then the long (positive) leads, avoiding any short-long contact. Solder a 15Ω resistor to the positive side (Figure B), and clip the excess length on all leads. Keep the whole affair small, with leads as short as possible, so it will all fit in the housing.

Add wires to connect to the switch and power (Figure C). Solder one to the negative leads and the other to the resistor, marking which one is which.

3. Put the light into the housing.

Fit the disk into the rubber end of the adapter, with the wires exiting the threaded hole in the back (Figure D). The rubber held my disk well without glue, but otherwise I would suggest a thin film of epoxy. I cut off just enough extra rubber to make a shim which, along with some electrical tape, holds the wires in back and keeps out water and debris.

4. Wire the circuit.

I mounted the light on top of my handlebar, the switch against the stem, and the battery pack behind the head tube. Trim the wires to the right lengths to connect these, leaving enough slack to let you turn the handlebar. The switch connects between the negative LED lead and negative battery terminal, and the positive LED lead connects to the red, positive battery terminal (Figure E).

5. Attach the light.

I mounted my light by interlocking the hose clamp that came with the adapter with a second clamp around the handlebar. For the switch and battery pack, I used zip ties, and added more to hold the wires against the frame (Figure F). Make it all tight so that nothing falls off if you go over a big bump.

Version 2

I made a second light for my girlfriend that has a better switch setup (page 129). I mounted a micro-mini switch in the hose adapter's hole in back, and routed the wires out a hole drilled through the side. This eliminates the big switch zip-tied to the handlebar.

+ You can see a wiring diagram at makezine.com/14/diycircuits_bikelight.

MIT student Trevor Shannon (trevorshp.com) has been making things since he was young. Occasionally, those things work.

EVASIVE BEEPING THING

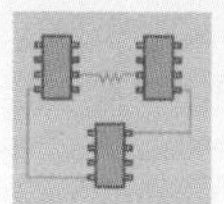

Infernal noisemaker turns pals into enemies. By Brad Graham and Kathy McGowan

Photograph by Sam Murphy

The Evasive Beeping Thing is appropriately named, since it dutifully does exactly what its name implies: it sends out a 5-second, high-pitched beep every few minutes. The source is extremely difficult to locate because of the way that high frequencies can penetrate objects and trick our ears.

You've probably encountered something similar in the real world, such as a failing appliance or a beeping wristwatch buried deep in a couch. As you know, high-pitched sounds seem like they are coming from all directions, which makes tracking them to the source a real chore. Add the fact that the sound only happens once every several minutes, and it may drive a person loopy as they spend all day looking for the source of the sound. Well, that's our goal, anyhow!

To generate the high-pitched audio wave, you'll connect a small speaker like those found in tiny electronic devices (cellphones, transistor radios, a tweeter from a small speaker system, etc.) to a simple audio oscillator set to a frequency near the upper limits of our hearing capabilities. The oscillator is triggered to run for approximately 5 seconds every few minutes by a 555 timer circuit with its output connected to the oscillator. The higher the frequency rating of the speaker, the farther the

MATERIALS

2N3904 transistors (2) **or any generic NPN style**
Resistors: 1MΩ, 100kΩ, 10kΩ (2), 1kΩ, and 100Ω
Capacitors: 100µF, 0.01µF
555 timer
Small speaker **from a transistor radio or the like**
9V battery and battery clip
Small perforated board **to wire the parts onto**
Container **to hide the unit inside**

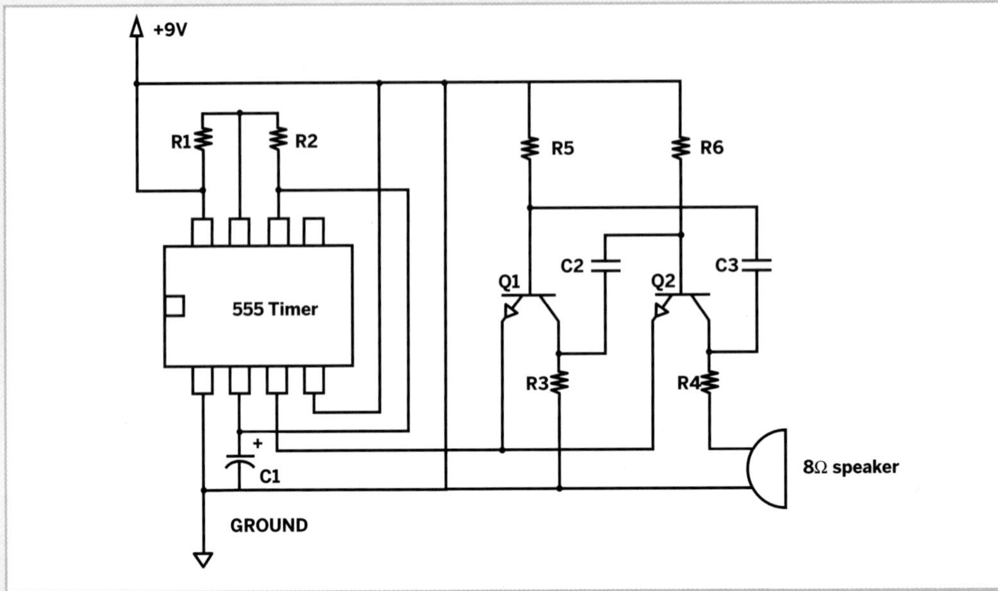

EVASIVE BEEPING THING SCHEMATIC: The 555 is set up so that its output will turn on the 2 transistor audio oscillators formed by the pair of NPN transistors.

high-pitched sound will travel, which is why a 2" or 3"-diameter tweeter is optimal for this project.

The small speakers shown in Figure B are perfect for this project. The rating of the small speaker is not important, since the audio oscillator will drive speakers from 4Ω to 16Ω with very little power output. The speaker on the top left was the one we decided to use in the final design because it fit nicely into the cabinet we chose to help disguise the evil device. Now, let's get on to the design of the electronics that make this unit work.

Above, you'll see the schematic of the Beeping Thing. The 555 is set up so that its output will turn on the 2 transistor audio oscillators formed by the pair of NPN transistors. Just like most 555 timer circuits, the timing cycle is controlled by the 2 resistors on pins 6, 7, and 8, and by the capacitor connected to pins 1 and 2. If you play around with the values of the 2 resistors, you can control the duty cycle of the timing pulses in order to alter both the off time and on time, to create more or less beeping each time the cycle repeats.

The capacitor controls the actual frequency of the timing pulses: the larger the value, the longer the duration between each timing cycle. In a really large room, you might want a longer beep and cycle, so a 220μF capacitor could be used, and the 100kΩ resistor could be swapped for a 220kΩ resistor. For a smaller room, where it may be easier to locate the device (e.g. a friend's office), the capacitor could be changed to 47μF and the 100kΩ resistor to a 10kΩ for a very short beep. The best plan is to simply build the unit as is, and then fine-tune the components until you're happy with its operation. And yes, a variable resistor would be easy to adjust.

Now, where do you hide the beast? Well, since this unit emits hard-to-locate high frequencies, your options are endless. The high-pitched sound will exit through the smallest hole in whatever box you place the parts into. We decided to cram the works into an old wall adapter (Figure A) that has all of the guts removed, including any connection to the AC lines. The little speaker fits nicely into the top of the box, and there's just enough room for the 9V battery and small circuit board. Figure C shows the completed circuit going into the wall wart box.

There was just enough room to get all the parts inside, so we couldn't install an on/off switch. But that was OK since the top of the box just snapped together and we could simply unclip the battery. The unit will run for many days on a full battery, and if you strategically place the beeper, it may take that

Fig. A: This annoying device eludes detection.
Fig. B: Several small, high-frequency speakers; the rating isn't important.

Fig. C: Installing the parts into an innocuous wall wart case makes it hard to find the source of the annoying beep!

long for the unsuspecting victim to find it! If you plan to use a wall wart cabinet for the device like we did, ensure that there's no connection between the plug prongs and the AC lines. It's a good idea to remove the prongs completely.

Some other good hiding places might be a pop can, lunch box, wall clock, tissue box, or even a working appliance. A solid cabinet will need a small hole for the speaker. We found that a ¼" hole was large enough for the tiny 2" speaker we used. You can also use a piezo buzzer instead of a speaker, which would make the unit even smaller and possibly louder owing to the very good high-pitched operation of the piezo element. To use a piezo buzzer in place of the speaker, connect resistor R4 (which used to connect to one of the speaker terminals) directly to the +9V line, where the other speaker terminal used to connect.

Now you can place the piezo buzzer in parallel with R4 to make it function. This is done because the piezo element will offer very high resistance as compared to the very low resistance of the speaker, and the current from the battery needs to flow to transistor Q2's collector.

The final product looks at home just about anywhere there is a wall socket, and can be easily hidden under furniture or inside another appliance for truly covert, mind-warping, annoying fun and games. We covered up the voltage switch from the original wall wart with black tape, and the little hole on the top of the case is barely large enough to pass a decent amount of high-pitched sound.

With the component values given, the beep emits about once every 3 minutes and lasts for approximately 5 seconds, just enough time to entice the victim to look for the source of the sound before it goes silent. We like to drop the unit in a room, then claim that we can't hear any beeping. This really gets the "beeper hunter" ticked off, and they try even harder to track down the evasive beeping thing to no avail. "I don't hear anything, pal, maybe you need an ear exam, or you should stop listening to pirated music on your MP3 player. I heard that the new copy protection can make your ears ring for days!"

Reprinted with permission from 51 High-Tech Practical Jokes for the Evil Genius *(McGraw-Hill/TAB Electronics), by Brad Graham and Kathy McGowan.*

Brad Graham is an author and inventor of electronics, custom bikes, robots, and Evil Genius works. He also hosts atomiczombie.com with his partner Kathy McGowan.

Photography by Brad Graham

DIY SCIENCE

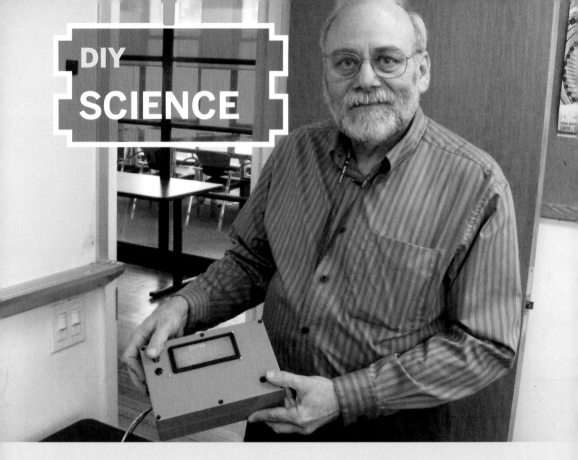

SAFETY SPECTROMETER

Device identifies dangerous liquids by analyzing light. By Eric Rosenthal

After air travel security banned bottled water and baby formula, I began wondering why they didn't use a device to determine the contents of liquids. If a liquid was detected to be safe, security could allow it on the plane. Spectrometers can identify the chemical makeup of a material by shining light on it and analyzing the precise mix of colors that bounce back.

These devices are usually very expensive, but I've designed a simple and inexpensive one that can identify liquids. You can also adapt it to determine the color of a swatch of paper or cloth or to identify a gem or semiprecious stone.

I spent less than $100 on this project and it took just a few days to design, fabricate, and test the hardware, plus another two days to write and debug the source code. Collecting the liquids and building the database took one evening, and it was fun!

MATERIALS

Arduino board from sparkfun.com. Use the Arduino NG or the latest USB version, the Arduino Diecimila.
LEDs (5) blue, green, yellow, red, and infrared
Infrared (IR) phototransistor
¼-watt resistors: 220Ω (5), 1KΩ (2), 2.2KΩ, 18KΩ
Serial display I used a Crystalfontz 634 Serial LCD; you could also use the Matrix Orbital LK 204-25, or similar products from seetron.com.
Power supply 6V–12V DC, 1A–1.5A
7805 5V voltage regulator and heat sink to drop the 12V to 5V for the display's backlight
Case from vellemanusa.com
Push-button switches (2) momentary, normally open
Soldering iron and solder
Wiring diagram Download from makezine.com/14/diyscience_spectrometer or follow the one in this article.

Photograph by Phillip Torrone

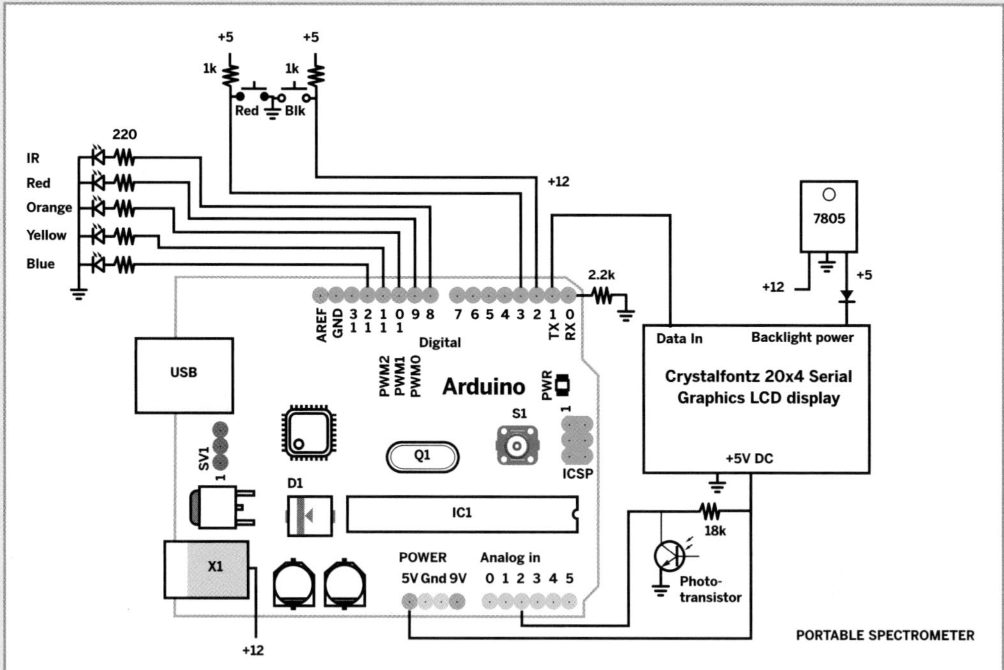

SCHEMATIC DIAGRAM OF SPECTROMETER: An all-controlling Arduino board drives 5 colored LEDs from output pins, along with a serial LCD display. Digital inputs from buttons switch between "learn" and "identify" modes, and analog input from the phototransistor is analyzed to identify the sample. A 7805 voltage regulator powers the LCD backlight.

A Little Science Background

A spectrometer measures the properties of light over a specific portion of the electromagnetic spectrum. Because all materials have a unique spectral signature, spectroscopic analysis can identify materials from the light that they reflect or emit. Mixtures of materials produce combined spectra, and by measuring the intensity of light at each wavelength, a spectrometer can determine the overall chemical makeup of a material under investigation.

When material burns, a spectrometer can detect and analyze the light it emits to determine the material's composition. In astronomy, highly specialized spectrometers are used to determine the composition of the gases that are ionizing and emitted as light energy from a star.

How It's Done

1. An Arduino board sequentially illuminates 5 different colored LEDs (light emitting diodes): blue, green, yellow, red, and infrared.

2. As each LED's light passes through a vial of liquid, we measure the intensity of the light detected by a phototransistor. See the wiring diagram above to wire together this part of the spectrometer.

3. Our spectrometer has a "learn" mode and an "identify" mode. In the learn mode, a known sample is placed in the unit and sampled at each wavelength emitted by the LEDs. The sampled values are stored in the Arduino memory. In the identify mode, an unknown sample is spectrally scanned, and the software in the Arduino compares the values of the scan with the values stored in the database.

4. A simple algorithm makes a best guess to identify the liquid, which is then displayed on a serial LCD. I think you'll be impressed by its accuracy.

Go online to get the wiring diagram, source code for the Arduino, parts list, and other info at creative-technology.net/MAKE.html.

See the spectrometer in action and get a peek under the hood at makezine.com/14/diyscience_spectrometer.

More photos at makezine.com/go/spectrometer.

Eric Rosenthal is president of Creative Technology, LLC (CTech), a company specializing in new and advanced imaging technology consulting and development.

DIY SPY

PARABOLIC MICROPHONE

This dollar store DIY spy mic lets you listen from afar. By Jim Lee

This is a ridiculously easy way to build a parabolic microphone using dollar store items. You'll attract lots of attention walking around in public with this rig. I usually welcome the inquiries, and let people listen to what I'm doing. Kids especially love it.

1. Make the dish.

Use wire cutters to snip away the 4 plastic holders that connect the hat's umbrella to its headband (Figure A). Slice the top of the plastic knob off the top of the umbrella (Figure B), and clean up the hole with a knife or reamer. Cover 1 gore of the umbrella near the center with a trapezoidal piece of the gaffer's tape. Cut a small X-shaped incision through the tape and umbrella; this will be the reinforced hole that the microphone wire will pass through (Figure C).

MATERIALS

Umbrella hat Make sure the umbrella part is vinyl, not fabric, which doesn't reflect sound as well.
9" paint roller handle
Small microphone I used RadioShack's discontinued Stereo Hands-Free Tie-Pin Microphone (#33-3028); but any decent small microphone will work. Dollar store purists can instead use earbud headphones as a microphone, or do surgery on a $1 hands-free cellphone headset. The sound won't be as good, but you'll have a true $3 parabolic mic.
Gaffer's tape or other tape with a rough surface
Cable ties

TOOLS

Hammer, wire cutters/side cutters, hobby knife and razor saw, permanent marker, reamer (optional), file (optional), laser pointer (optional)

Photography by Jim Lee

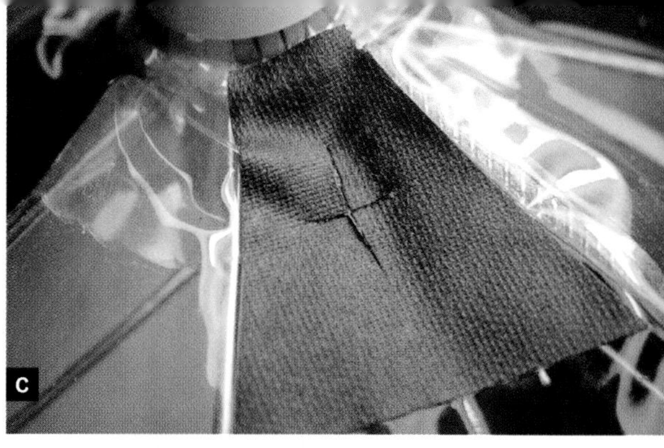

Fig. A: Use clippers to remove the plastic straps connecting the hat to the umbrella. Fig. B: Cut the tip off the umbrella knob. Fig. C: Cut an X into the umbrella to feed the microphone wire through.

Fig. D: Insert the paint roller rod through the tip of the umbrella. Fig. E: Install the microphone, then secure the cable to the paint roller with cable ties. Fig. F: Ready for spying.

2. Attach the handle.

Remove the paint roller's plastic caps and wire frame. Push the shaft through the hole in the top of the umbrella, so that it protrudes 6" underneath. Leave ½" of clearance between the outer surface of the umbrella and the bend of the handle (Figure D).

Just above the umbrella's top knob, wrap a length of tape around the shaft and ring it with a cable tie pulled tight. Wrap the the shaft with more tape, to provide a gripping surface for the microphone.

3. Install the microphone.

Clip the mic to the shaft and thread the cable through the X hole. Secure the cable with cable ties (Figure E).

You want to place the microphone at the focal point of the reflector, but realize that this is a plastic umbrella, not a perfect parabola. So this "point" will be more of a semifocal blur. Here are 3 ways to position the mic, in decreasing order of complexity:

3a. Point a laser at different points on the inside of the umbrella from a distance of about 20 feet directly in front of the unit. Mark where it reflects onto the shaft to find the general region of focus.

3b. Plug the mic into a recording device, put on some headphones, and point it toward a ticking clock some distance away. Move the microphone along the shaft until you get the loudest sound.

3c. Just take my word for it, and position the mic about 3" from the inside surface of the umbrella.

4. Take it for a test ride.

Plug your new parabolic mic (Figure F) into a recorder. Use headphones to monitor your work. Then point it at something interesting. You're in for a pleasant surprise! Now try recording the same sound without the parabolic setup — forget it.

♪ Hear field recordings of a squirrel and a cardinal made with the Dollar Store Parabolic Mic at makezine.com/14/diyspy_mic.

Jim Lee (bambooturtle.com) is an artist who lives in Durham, N.C., with hundreds of turtle and tortoise artworks, plus a few live ones.

DIY SPY

COVERT SPY SUNGLASSES

Record what you see and hear with these low-cost stealthy sunglasses. By Kip Kedersha

Ever since I was a kid I've loved spy gadgets. From *Wild Wild West* to *Get Smart* and *Mission Impossible*, they have always fascinated me. I can remember sitting in front of the tube watching the hidden camera pranks of *Candid Camera*. Well, now I have my own pair of sneaky sunglasses that are cheap and easy to make.

1. Route camera's cable to 1 side.

The first mod you may have to make is to route the wiring coming from the micro camera (Figure A). You want the wires to exit the camera on 1 side of the sunglasses instead of the top or bottom. Unscrew the camera housing and reroute the cable so that when the camera image is normal, the cable is routed as close as possible to the sunglasses temple. Either temple will work; I chose the right because I'm right-handed and it just seemed normal to have it on that side. Use the Dremel tool to cut a small notch in the housing so the cable passes through to the side you choose (Figure B).

2. Add a 4' extension cable.

Next you need to extend the camera cable so that it can run the length of the temple and then down your back or side to be concealed.

First, cut the micro camera's cable a few inches from the camera (Figure C). You always want to leave a little extra wire just in case you mess up. You should find 4 wires inside the insulation: red, yellow, black, and white, corresponding with the connectors on the end. If it's got the standard color-coding like mine does, red will be power, yellow video, white audio, and black the common ground for all the

Photography by Kip Kedersha

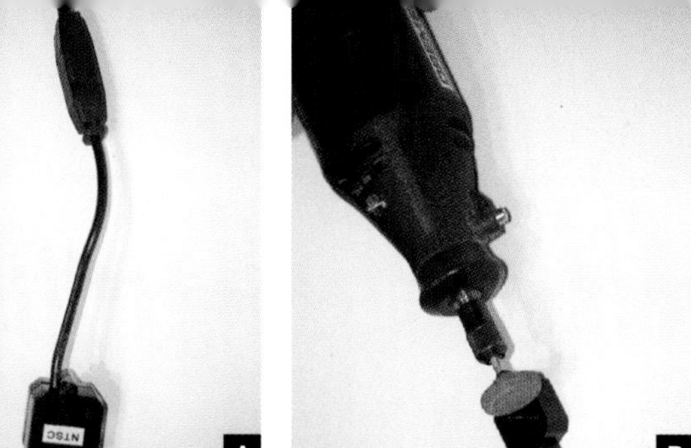

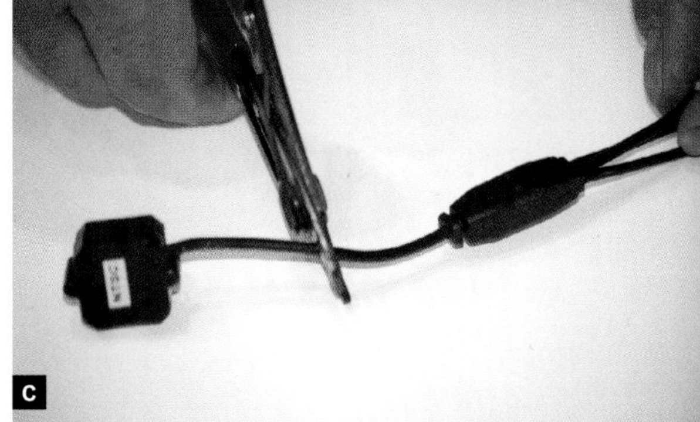

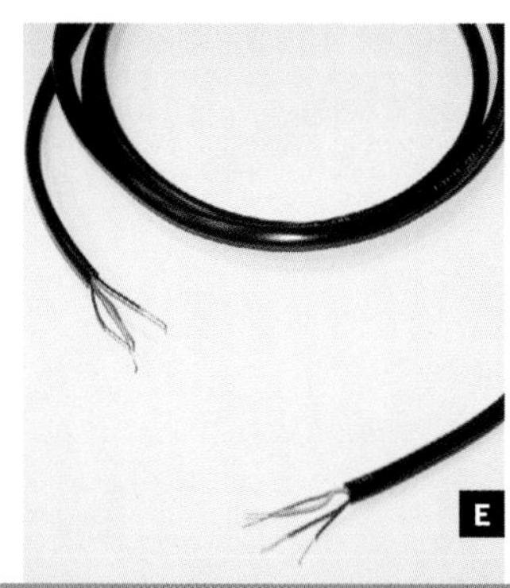

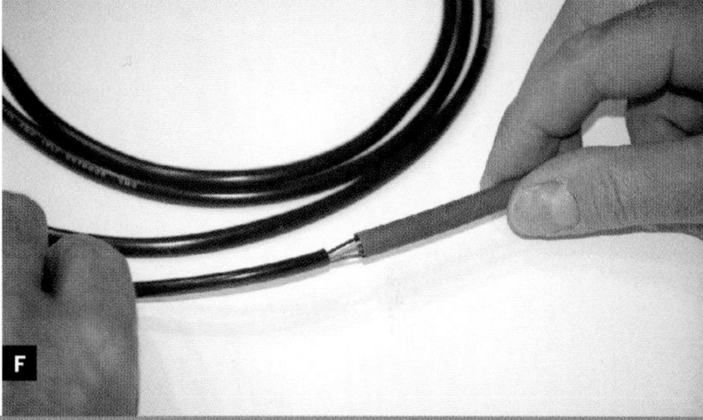

Fig. A: The micro camera you'll be using for this project. Fig. B: Use a Dremel tool to cut a routing notch in the camera housing. Fig. C: Cut the cable a few inches from the camera.

Fig. D: Strip wires on both ends. Fig. E: Make an extension cable and strip both ends. Fig. F: Slip heat-shrink tubing over both ends of the extension cable.

MATERIALS

- **Solar Shield sunglasses** or any brand in the wraparound style. I found mine for $19.
- **Micro color spy camera that records audio** You need one that can be powered by a 9V battery. eBay has the best selection; the camera I bought was $20 delivered to my door.
- **9V battery**
- **4-conductor wire, about 4'** This can be any small wire no larger than the diameter of the temple of the sunglasses. I used 24-gauge telephone wire.
- **Small camcorder** any brand with A/V inputs, small enough to fit in a coat or pants pocket. Mine is a Sony DCR-HC32.

TOOLS

- Wire stripper/cutter
- Soldering iron and solder
- Black heat-shrink tubing
- Hot glue gun
- Small screwdriver(s)
- X-Acto knife
- Electrical tape
- Dremel tool

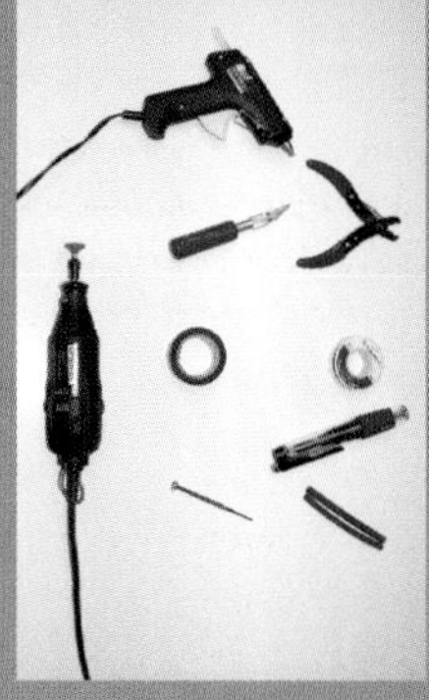

connectors. Cut back the insulation and strip the wires on both halves of the cable (Figure D).

Now prepare your extension cable. Cut a 4' length of 4-conductor wire, then trim the insulation back and strip the wires on both ends (Figure E). Cut 2 pieces of heat-shrink tubing and slip them onto the extension cable (Figure F). Twist all the connections together, taking note of the color-coding. Then attach your 9V battery to the red (power) connector, and check to make sure you're getting video and audio by hooking the cable up to your camcorder or to a TV monitor.

Once you're sure everything is wired correctly, solder the 4 wires from the camera to the extension cable (Figures G and H). I added a small piece of electrical tape to insulate the 4 connections from one another. Then slip the heat-shrink tubing over your connections and use a heat gun, hair dryer, or lighter to heat up the tubing until it becomes snug (Figure I). If you use a lighter, quickly move it back and forth under the tubing so you don't burn it. Now repeat the process with the other end of the cable.

3. Mount camera in the sunglasses.

Clean the sunglasses lens on the side where you'll install the camera, because once it's glued in place

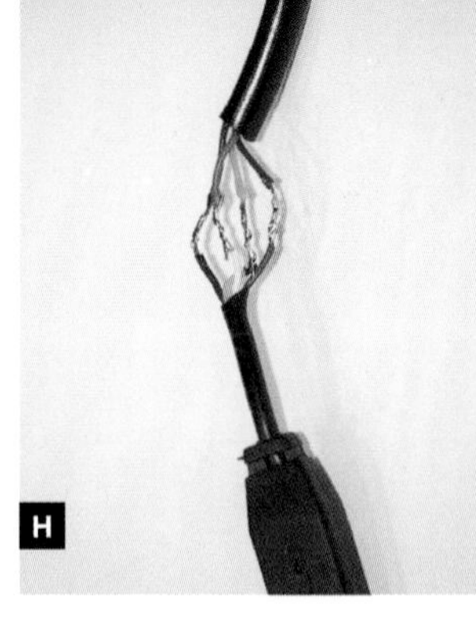

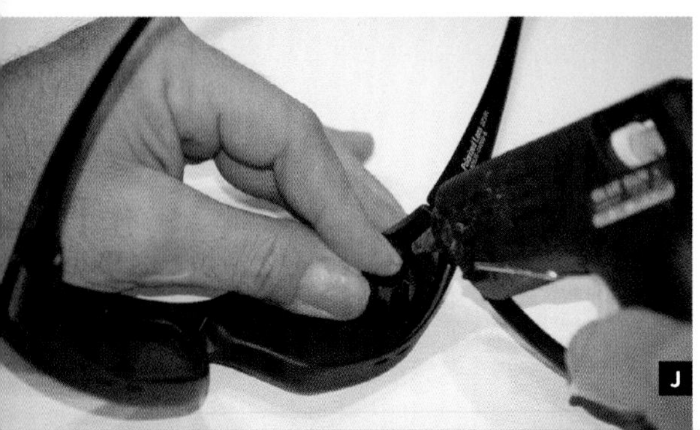

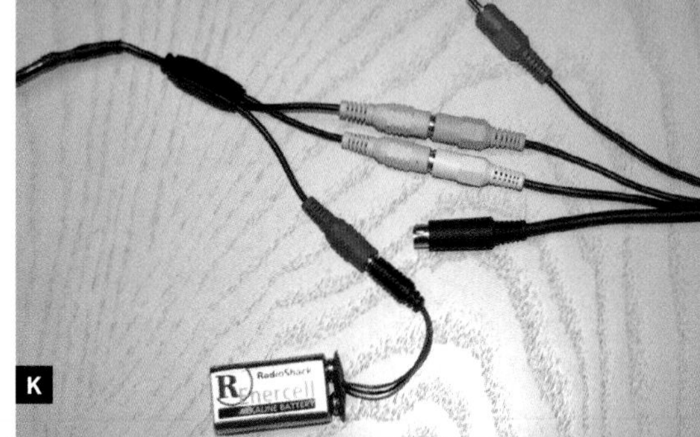

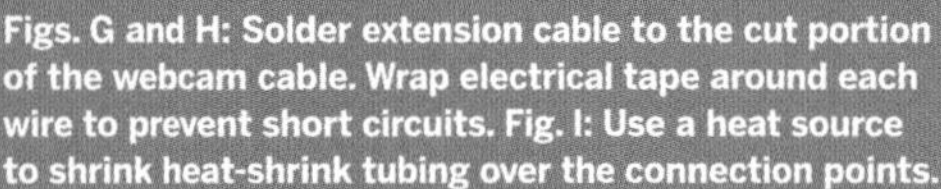

Figs. G and H: Solder extension cable to the cut portion of the webcam cable. Wrap electrical tape around each wire to prevent short circuits. Fig. I: Use a heat source to shrink heat-shrink tubing over the connection points.

Fig. J: Use hot glue to secure the webcam to the sunglasses. Hold the camera in place until the glue solidifies completely. Fig. K: Plug the cable into the camcorder and attach the 9V battery.

you won't have access to the lens in that area. Hook the camera up to a TV or camcorder again, position it in the far corner of the sunglasses, and check its positioning: it should shoot straight ahead and should be level with the ground when you're wearing the glasses. When you're sure of the location, heat up your glue gun and, while holding the camera in place, shoot an ample amount of glue in the corner to secure the camera (Figure J). Hold it in place for a few minutes until the glue is completely dry.

4. Test-fit the glasses and hide the cable and camcorder.

Now it's time to try on the glasses and see if you need to make any other adjustments. Depending on how much hair you have and the type of clothes you'll be wearing, the cable may just run behind your ear and down your back (see page 138). You can use a small amount of hot glue to secure the cable to the temple, or use a small piece of electrical tape if that will conceal it better. Another option is to use a pair of eyeglass straps or retainers and then hide the cable along those.

Get your camcorder out, plug in your extension cable, and attach your 9V battery. In my case, I had to use an A/V cable that plugged into the camcorder, then plug the extension cable from the micro camera into that A/V cable (Figure K). My camcorder is small enough to fit in a coat pocket or even the front pocket of a pair of jeans.

5. Spy and videotape!

Test it out! Go outside and walk around. If the camera is lined up correctly, whatever you look at will be recorded. Most of the micro cameras I've used have an auto iris that works great — on the video you won't even be able to tell that it was shooting through dark sunglasses.

Now come up with some creative uses for your spy specs. Do your own *Candid Camera*-style pranks, or take them with you to places where videotaping may be frowned upon. Don't break the law. Keep it safe and fun, and enjoy your new covert spy sunglasses!

Watch Kipkay's how-to video for the Covert Spy Sunglasses at makezine.com/go/spysunglasses.

Kip "Kipkay" Kedersha is a videographer with 25 years of video production experience. In his spare time, he makes videos that reflect his personal interests in DIY solutions to everyday problems. His work can be seen at kipkay.com.

DIY MUSIC

THE MACHINIST'S PHONOGRAPH

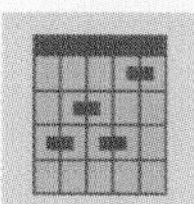

This time-tripping player handles all cylinder record formats. By Royston Maybery

When Edison invented the phonograph in 1877, he envisioned it as a business dictation machine. But it soon became a popular medium for music, recorded onto durable cylinders that are still available and playable today. Cylinders came in a wide variety of formats — 2¼", 3½", or 5" in diameter, 4" or 6" long, 120rpm–160rpm, 100 or 200 grooves per inch — most of which required different types of phonographs to play.

Today there's one specialty machine, the Archeophone, that can handle all formats of cylinder recordings, but it costs more than $16,000. So I decided to build my own, and being a licensed machinist, I call it the Machinist's Phonograph.

Design

Physically speaking, cylinder phonographs need to do 2 things: turn the cylinder record itself, and turn the screw that leads the tone arm along the cylinder's spiral groove at the proper speed. My phonograph uses a double belt drive to accomplish both tasks. One belt runs from the motor to the shaft carrying the cylinder, and a second belt connects that shaft to another one, which turns the lead screw.

The challenge was figuring out how to accommodate all cylinder formats. For the different speeds, I calibrated the motor pulley diameters to give the drive shaft a maximum speed of 170rpm,

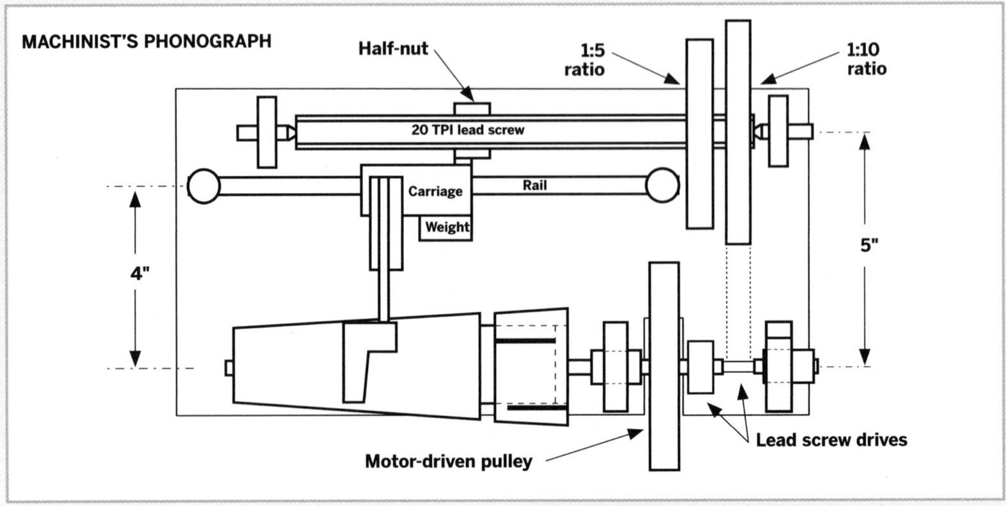

MATERIALS

- 3/8" aluminum plate, 12"×6¼"
- Aluminum blocks, 4"×1"×½" (4)
- ½" aluminum rods, 4" long (2)
- ¼"×20 cap screws (10)
- ¼" steel rods/dowels, 1" long (2)
- ½"×20 UN threaded rod, 8½" long
- 3/8" steel rod, 8½" long
- 5/16" steel rod, 9¾" long
- 3/8" ID Oilite bushings (2) **Made out of oil-impregnated bronze, these are self-lubricating.**
- Nylon rounds and scrap pieces
- Rubber drive belts, style U/7 (2) **for vacuum cleaner carpet brush**
- ¾"-thick pine board
- Wood screws
- Grub screws, ¼"×20 (7)
- AC fan motor **Mine was rated at 1,550rpm.**
- Fan speed control **I used an off-the-shelf wall control for a ceiling fan.**
- Wood glue, super glue, filler, stain, and varnish
- Tone arm, cartridge, and rubber feet from an old stereo turntable
- Sapphire stylus **for 78rpm records**
- Glass shards **to make a glass ball stylus**

TOOLS

- Vertical milling machine
- Small metalworking lathe with compound slide
- Tap and die set, ¼"×20 and ½"×20 UN **I tried to use as few taps as possible and allow the machine to be dismantled and assembled with as few tools as possible.**
- Table saw **or back saw and wood chisels**
- Drill and 3/8" reamer
- Sandpaper
- Gas flame and tweezers **to make glass ball stylus**

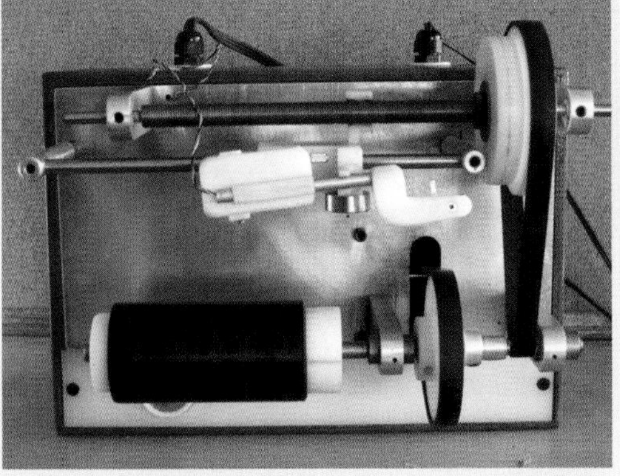

and included a knob that adjusts the speed down. For cylinders of different sizes, I made swappable mandrels to fit the drive shaft in the 3 standard diameters plus a longer mandrel to handle the 6"-long Columbia cylinders, along with matching sleeves that fit around the tone arm stalk to lift it to the proper height. The lead screw has 6½" of usable travel, more than enough to handle the job.

The toughest variant was groove pitch. Two-minute cylinders have 100 grooves per inch, and 4-minute cylinders have 200 per inch. To support both, the belt between the drive shaft and the lead screw has 2 positions over 2 adjacent pairs of pulleys, one with a 1:5 drive ratio and the other with 1:10. To change the speed, you move the rubber belt from one pair to the other. The belt material is sufficiently stretchable that I didn't need additional tensioner pulleys, and the tone arm tracks the groove accurately at both speeds. Any drift due to belt slippage or other machining factors is compensated for by the tone arm's ability to pivot laterally.

Metal Parts

I made most of the parts out of aluminum, steel rod, or nylon, using a small lathe and milling machine. Complete plans are at makezine.com/14/diymusic_phonograph. First, I made the base plate out of ⅜"-thick aluminum. I indexed the mounting holes using the digital readout of my milling machine because it was available, but that was probably overkill; a skilled operator could do it manually with a ruler, scriber, punch, and drill press.

Then I made the 6 uprights to hold the drive shaft, lead screw, and carriage rail. These are identical pairs of aluminum blocks or rods machined and tapped in various ways (see the plans) and mounted to the base plate with ¼"×20 cap screws. I also turned two 1" pieces of ¼" steel dowel in a lathe to sharpen them to 60° points, which fit into the lead screw uprights. The points hold the screw horizontally so that it can rotate.

The drive shaft, lead screw, and carriage rail themselves are made from standard steel rods. The carriage rail is simply a 5⁄16" steel rod cut to length. The lead screw is ½"×20 threaded rod, faced and center-drilled at both ends with a dab of grease for the points. The drive shaft is ⅜" steel rod, held in its uprights by a pair of Oilite bushings. I used a lathe to turn a ¾" section of the rod near 1 end into a crowned (rounded) drive pulley, for the 1:10 drive configuration. Flat belts stay on crowned pulleys because they gravitate to the highest point.

The phonograph also has 2 crowned drive pulley pieces made of aluminum, one for the motor underneath to power the drive shaft, and the other for the drive shaft to power the lead screw in the 1:5 configuration. Both are held in place by grub screws through their flanges.

Nylon Parts

I used nylon to make other components. I turned the phonograph's 3 driven pulleys on the lathe. The motor-driven pulley has a flange for a grub screw to secure it to the drive shaft. The 2 driven pulleys on the lead screw are simple disks, secured on either side by ½"×20 nuts, thinned down on the lathe, and washers; 2 more washers in between the pulleys act as spacers.

Opposite the pulleys on the drive shaft is the tapered nylon mandrel that holds the cylinders, along with a sliding, tapered sleeve that helps keep them snug. I also made both of these on the lathe, using the compound slide to attain the angle on the sleeve and offsetting the tail stock to achieve the angle on the mandrel.

For the tone arm carriage, I cut a block of nylon and drilled 2 parallel holes along its length 1" apart: a smooth 8mm hole that slides over the rail, and a ½"×20 threaded hole that engages with the lead screw, which drives it. Then I cut away the top half of the threaded hole to create a half-nut, so you can disengage the carriage from the screw and slide it back to the beginning of the cylinder. A metal weight on the cylinder side of the carriage acts as a counterweight to push the half-nut up against the lead screw, with the rail in between acting as the fulcrum (Figure A, following page).

Box, Motor, and Speed Control

The box for the phonograph I made out of pine, cutting the pieces on a table saw, joining them with screws and glue, and finishing with some wood filler followed by a sanding and a lick of stain and varnish.

I screwed the fan motor to the underside of the base plate, which simply rests on the inside lip of the box. I made the lip with a couple of passes on the table saw (Figures B and C, following page).

A long strip of nylon screwed to the box runs flat alongside the base plate. This carries the phonograph's speed control, which is an off-the-shelf wall control for a ceiling fan. On the back side of the box, 2 nylon donuts reinforce the holes for the power and (mono) audio cords.

Tone Arm, Cartridge, and Needles

My tone arm came from an old BSR stereo turntable, and I rewired the cartridge to read cylinder recordings. The challenge is that cylinder recordings are vertical, where the groove varies in depth. In contrast, mono 78rpm and 45rpm discs are lateral, with a zigzagging groove, and 33rpm stereo LPs combine both approaches. To convert a stereo cartridge to read vertical tracking as its main signal, reverse the wires on either the right or the left channel. If that doesn't work, reverse them on the other side. The result should be clear sound from both speakers.

To play 4-minute cylinders, the cartridge holds a sapphire stylus for 78rpm records.

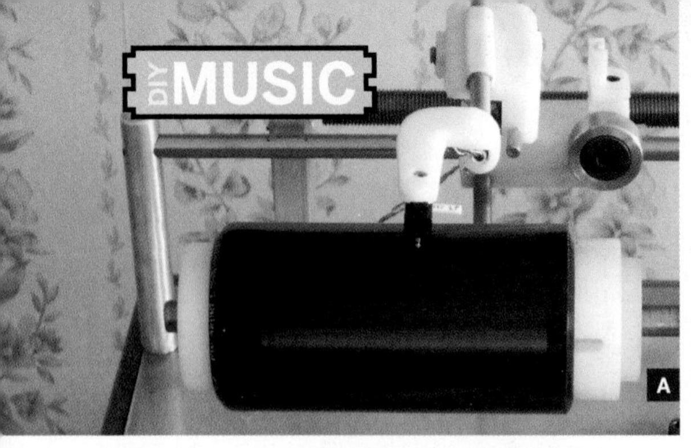

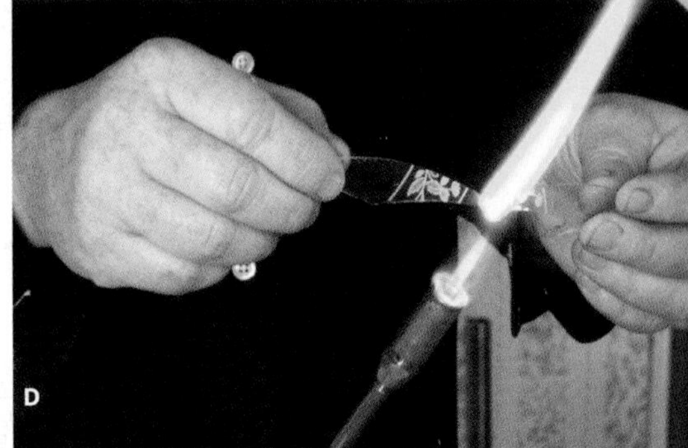

Fig. A: A donut-shaped metal weight pulls the carriage into position and lifts the half-nut against the drive screw in back. Fig. B: The lid, opened to show the motor and both drive belts.

Fig. C: The wiring on the underside of the lid for the fan motor control and stylus. Fig. D: Glass thread is pulled from shards to make the stylus.

NOTE: Don't use a diamond stylus; these will damage cylinder recordings.

For 2-minute cylinders, which are softer, I used a homemade glass ball stylus, following the technique described by Rob Lomas on the Phonograph Makers' Pages (christerhamp.se/phono).

To do this, take 2 pieces of broken glass (I used wine glass shards), and hold their pointed ends over a gas flame. When they both glow bright red, touch them together to fuse them, then quickly pull the them apart to create a glass thread about 2½" long (Figure D). The thread should be about 0.002" thick, and it may take a few tries to achieve. Break the thread into ½" lengths, and use tweezers to hold 1 end over the flame until a small bead forms, about 0.004" in diameter. Super-glue the glass stub, ball side down, to the point of an old stylus body.

Download complete plans and hear cylinder recordings played by the Machinist's Phonograph at makezine.com/14/diymusic_phonograph.

Royston Maybery is a licensed machinist and registered teacher in Ontario, Canada, and holds a master's degree in history from Sheffield Hallam University in the U.K.

CYLINDER RECORDINGS

The invention of the disc record in 1887 spelled the end of cylinder recordings, but cylinders were still manufactured up until the stock market crash in 1929. Today they have a following as collectibles. The recordings offer a view into the musical tastes and values from the era of our great-grandparents, and are a valuable primary source for historians. You can find them on eBay and in flea markets and antique shops.

The most common content was music, both classical and popular. Popular music tended to be simple sentimental songs, military bands, patriotic tunes, ragtime, spirituals, and songs that would be considered racist and offensive by today's standards, yet express the sentiments of their time. Sermons, comedy routines (also frequently racist), and political speeches also appear, some of the latter by notables such as presidents William Howard Taft and Theodore Roosevelt. You can learn more from collectors clubs and other resources listed at makezine.com/14/diymusic_phonograph.

NO POSTAGE
NECESSARY
IF MAILED
IN THE
UNITED STATES

BUSINESS REPLY MAIL

FIRST-CLASS MAIL PERMIT NO 865 NORTH HOLLYWOOD CA

POSTAGE WILL BE PAID BY ADDRESSEE

Make:

PO BOX 17046
NORTH HOLLYWOOD CA 91615-9588

NO POSTAGE
NECESSARY
IF MAILED
IN THE
UNITED STATES

BUSINESS REPLY MAIL

FIRST-CLASS MAIL PERMIT NO 865 NORTH HOLLYWOOD CA

POSTAGE WILL BE PAID BY ADDRESSEE

Make:

PO BOX 17046
NORTH HOLLYWOOD CA 91615-9588

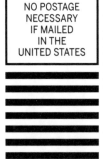

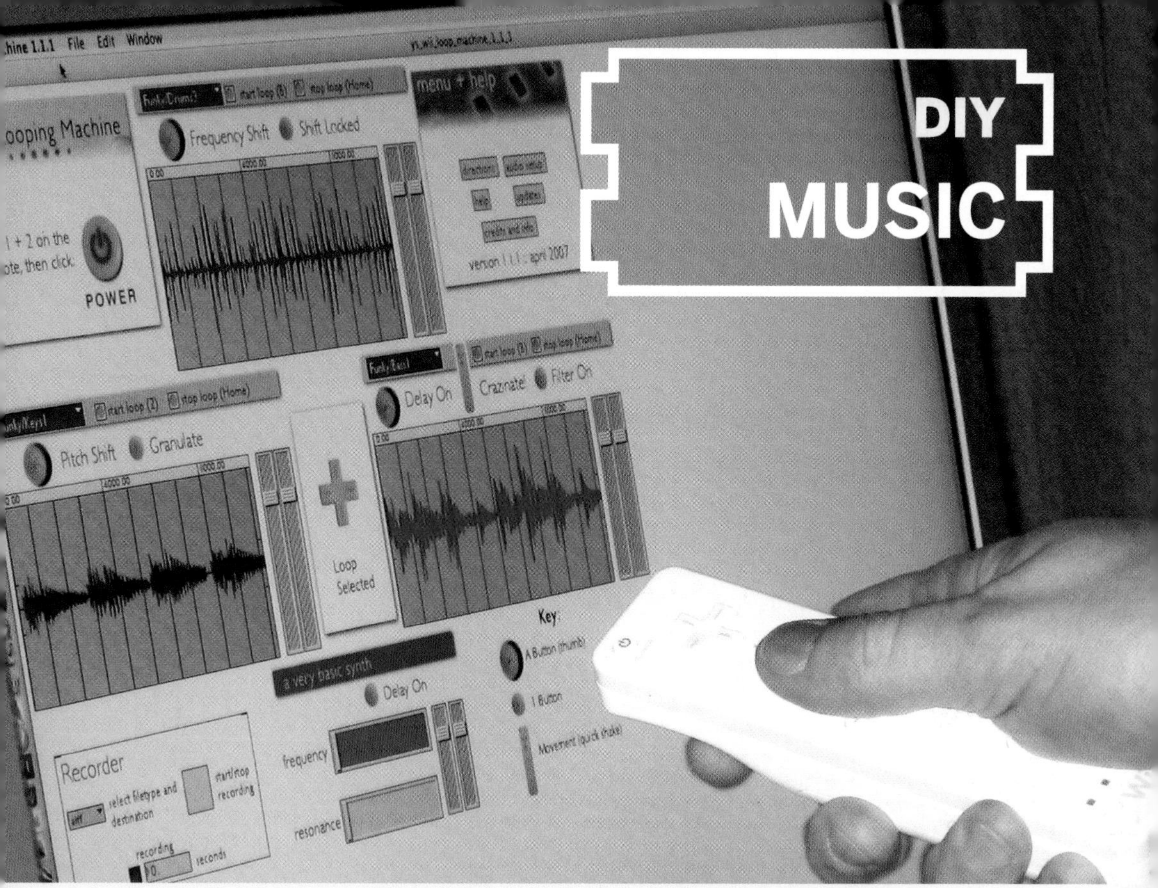

WII WILL ROCK YOU

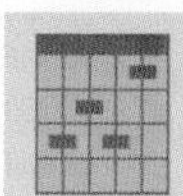

Play real air guitar (or keys) with the wireless game remote. By Bill Byrne

I grew up in the 80s with a Commodore 64, Game Boy, and Nintendo Entertainment System, and I feel more comfortable with an NES game pad than a TV remote.

Many years later, the NES game pad has evolved into a monster: the Wii Remote wireless game controller, affectionately known as the Wiimote. My wife, Suzanne, and I also perform and record electronic music in a band called the Painful Leg Injuries. I've played with numerous MIDI controllers, but despite my years of childhood piano lessons, nothing feels as natural to me as the Wiimote. Blame it on all those hours with *Tetris* and Mario.

Here, I've written up 4 of the setups I'm using to control music with this thereminesque, ether-bending joystick.

Photograph by Bill Byrne

1. Loop Machine

HARDWARE

Mac laptop, Bluetooth wireless-enabled

Nintendo Wii Remote controller **just the controller, you don't need the full Wii system**

Wii Nunchuk attachment (optional) **well worth it**

SOFTWARE

Loop Machine version 2.0 **$20, or Wii Loop Machine version 1.1.1, free, from** theamazingrolo.net/wii

Yann Seznec's Loop Machine (Figure A, following page) is a great way to start warping sounds. It has 3 channels you can load audio samples into, and an additional channel that functions as a basic synth.

The Wiimote's buttons switch among the channels

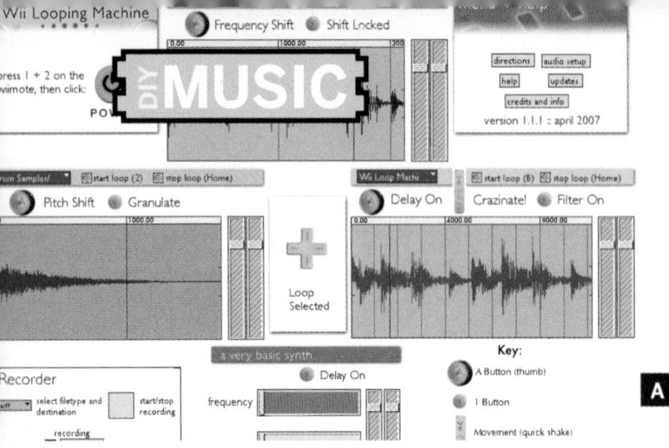

The Wiinstrument

BETA

DrumStiicks (MIDI)

Smack your Wii remote and Nunchuk like drum sticks to play MIDI notes.

Press a selection button to assign a played note on Wiimote hits.

Hold B simultaneously to assign the note to the Nunchuk.

Use Left and Right to shift the scale by a halftone.

Up C-1
Down C#-1
A D-1
- D#-1
+ E-1
1 F-1
2 F#-1

B

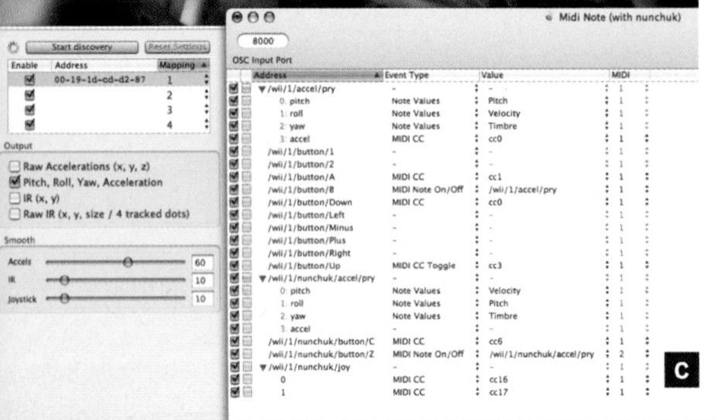

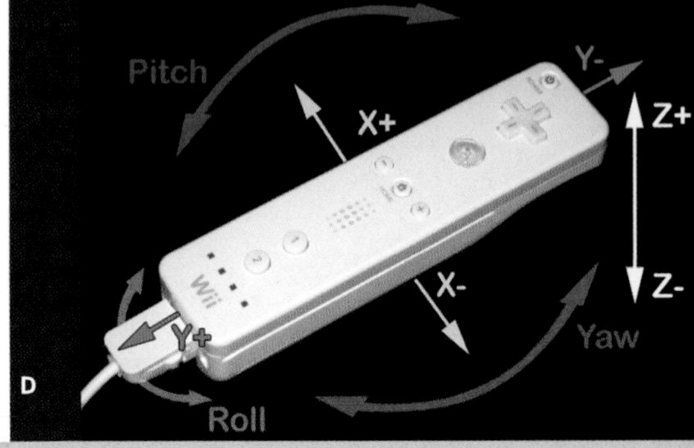

Fig. A: Loop Machine is a great way to get started Wiimote-controlling sounds. Fig. B: The Wiinstrument lets your Wiimote speak MIDI.

Fig. C: The OSCulator Wii-MIDI interface can associate any Wiimote actions to any MIDI commands. Fig. D: Wiimote motions: roll, pitch, and yaw.

and let you trigger, loop, and apply effects to the audio for each. Moving the Wiimote around tweaks the effects settings to change the sound according to your movements. For a track my band recently recorded, I loaded some of Michael Winslow's vocal beatbox sounds from the *Police Academy* movies into all 3 sample channels. I created a bass loop using 1 channel, then pitch-shifted and tweaked the sample in the other 2 to overlay it with weird, out-of-phase variations.

2. The Wiinstrument

HARDWARE
Same as previous

SOFTWARE
The Wiinstrument free from screenfashion.org
GarageBand or any other digital audio workstation (DAW) or soft-synth. GarageBand is part of the iLife package, $79 from apple.com/ilife.

Loop Machine is fun, but it's just the tip of the iceberg. Instead of working as a simple stand-alone instrument, Screen Fashion's The Wiinstrument (Figure B) serves mainly as a Wii-MIDI port, although it also plays drum samples. This means it can plug your Wiimote into any digital audio workstation (DAW) or MIDI soft-synth. I paired The Wiinstrument with GarageBand, and it turned my Wiimote and Nunchuk into a nice drum controller. One caveat: if you follow Screen Fashion's GarageBand tutorial, you may have to turn on the Compressor effect and raise its Threshold and Gain settings; otherwise when using Drumstiicks mode, the volume can be very low.

3. OSCulator

HARDWARE
Same as previous

SOFTWARE
OSCulator 2.5.6 $19 minimum donation, osculator.net
Ableton Live or any other DAW or soft-synth. Ableton Live LE is $149 and the Ableton Live Suite is $499 from ableton.com.

The Wiinstrument doesn't allow you to save your settings, and the more you play your Wiimote, the more you'll want to customize its control setup.

Wii Controller for Outboard MIDI Gear

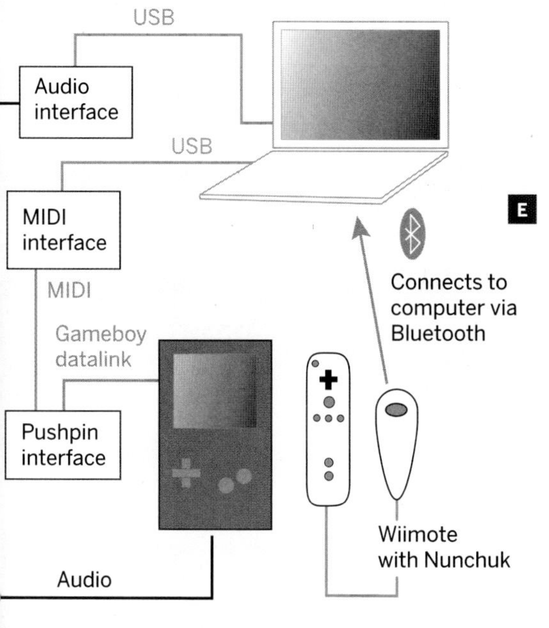

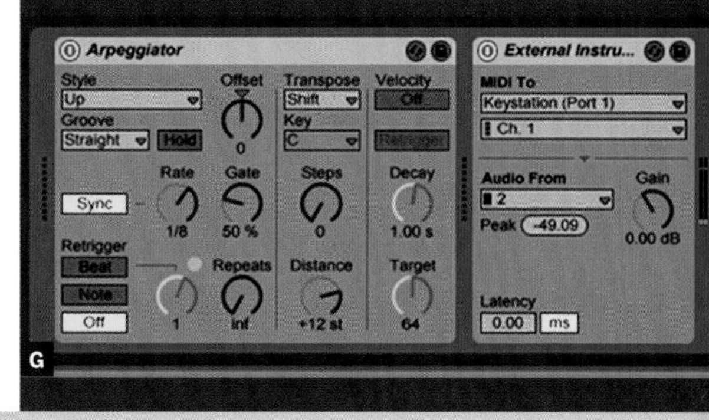

Fig. E: Hardware setup for Wiimote-controlling outboard Game Boy Color. Wiimote generates MIDI commands via OSCulator, which run out to Game Boy. Audio runs from the Game Boy headphone jack back to the laptop.

Fig. F: This breadboarded Pushpin hardware interface connects MIDI input to the Game Boy link cable.
Fig. G: Ableton Live's Arpeggiator plugin is great for creating 8-bit video-game-sounding melodies.

That's where OSCulator comes in.

OSCulator (Figure C) is the ultimate tweakable gateway between your Wiimote and MIDI-controlled audio. It lets you set up exactly how you want the Wiimote to react to each action, either button pushes or rotations along its 3 axes: roll, pitch, and yaw (Figure D). You can assign a button to send Note On/Off commands (I like using the B trigger at the bottom for this) and assign which rotation determines the note's pitch once it's fired.

You can also set MIDI CC (Continuous Control) commands to various buttons and rotations, to vary volume, reverb level, and other note properties. You can research which MIDI CC number controls what property, and it's also fun to just experiment and listen to what happens.

To get started, load one of OSCulator's 2 preset patches for Live, Midi Note or Midi Note with Nunchuk, and select OSCulator In as an input in the MIDI preferences. Then set up a MIDI track with a Sampler as its instrument, and drop in a short sample. Move your Wiimote around while holding down the B trigger. A note will play when the button is pressed, and your movements control the pitch. For my band's track "Hirsute Head Hunters Hole-Hearted Happiness," I used this setting to trigger samples of my trumpet playing, in an attempt to conjure up *On the Corner*-era Miles Davis played by a broken robot.

OSCulator's Midi Note with Nunchuk preset is designed for Live, but I also love using it to play drums in GarageBand; it creates a very liquid sound, which I used throughout our most recent album.

4. Outboard MIDI Hardware

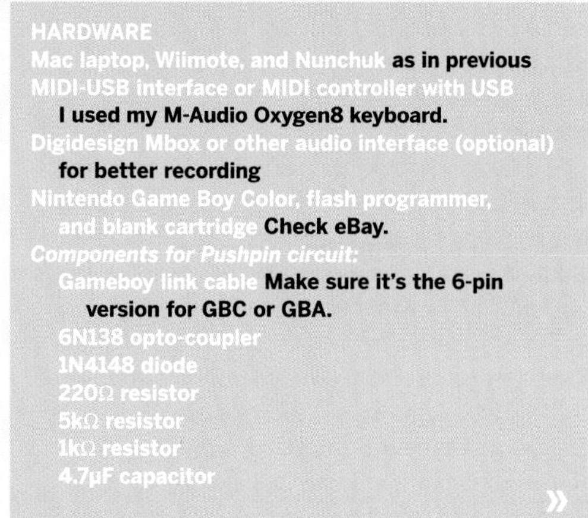
HARDWARE
Mac laptop, Wiimote, and Nunchuk **as in previous**
MIDI-USB interface or MIDI controller with USB **I used my M-Audio Oxygen8 keyboard.**
Digidesign Mbox or other audio interface (optional) **for better recording**
Nintendo Game Boy Color, flash programmer, and blank cartridge **Check eBay.**
Components for Pushpin circuit:
Gameboy link cable **Make sure it's the 6-pin version for GBC or GBA.**
6N138 opto-coupler
1N4148 diode
220Ω resistor
5kΩ resistor
1kΩ resistor
4.7μF capacitor
»

Photography by Bill Byrne

5-pin DIN MIDI jack
Solderless breadboard/perfboard **For keeps, you can solder onto perfboard later. See makezine.com/go/pushpin for details.**

SOFTWARE
Pushpin MIDI Music Synthesizer ROM binary **free from makezine.com/go/variogram**
OSCulator and Ableton Live **as in previous**

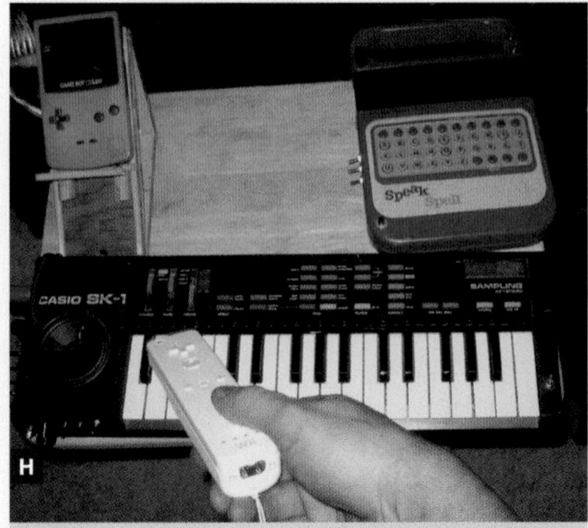

Fig. H: Wiimote-controlling a Casio SK-1 keyboard and a Speak & Spell toy.

My next step was to Wiimote-control some MIDI outboard gear. There's a DIY package called Pushpin that retrofits a Game Boy Color to work as a synthesizer, and I was excited to try it out and see if I could run it with my Wiimote — at first for the anachronistic-irony appeal, but that was soon overshadowed when I heard how amazing Pushpin sounds.

Figure E, on the previous page, shows the hardware setup. To take the Wiimote's MIDI from OSCulator and patch it over to the USB port that feeds the Game Boy Color, I used Ableton Live's External Instrument plugin. This can run through a plain USB-MIDI interface, but I used a keyboard controller to add more playing options.

To get Pushpin going, you need to assemble some electronics that interface between the MIDI cable and a hacked Game Boy link cable (Figure F, previous page). This feeds the Game Boy running the Pushpin software, which in turn sends synthesized audio out the headphone jack. The Pushpin website (makezine.com/go/variogram) has instructions for building the circuit and flashing the cartridge.

I then fed the audio back to Ableton Live on my laptop through a Digidesign Mbox audio interface. This is optional; the Mac can actually do a lot with audio coming directly into its microphone jack, but using an outboard interface like the Mbox lets you mix levels and otherwise gain more control over the signal.

I've fallen in love with using Live's Arpeggiator plugin (Figure G, previous page) to get classic video-game-sounding melodies. For a track devoted to Game Boy sounds, I recorded the Game Boy's Pushpin audio along with the source MIDI that generated it. Then I copied the MIDI data to another track in Live, changed the instrument to Simpler, and used the MIDI to trigger sound samples from LSDJ, the classic old Game Boy music cartridge.

In addition to the Game Boy Color, I've also experimented with Wiimote-controlling a Casio SK-1 keyboard and a Speak & Spell toy (Figure H). I used kits from Highly Liquid (highlyliquid.com) to retrofit these two circuit-bending favorites as MIDI devices, then I connected them into the same setup as the Game Boy Color and Pushpin interface.

Standard MIDI keyboard controllers give you piano keys and a pitch wheel, which may be great for Mozart or Billy Joel, but the Wiimote lets you draw a note's pitch in the immediate space in front of you. Making music with the Wiimote feels painterly.

Even if you are a seasoned MIDI keyboardist, the Wiimote music experience is worth the effort and cost. It's a blast, and also a great way for nonmusicians to enter the world of digital music. Especially if you hated those piano lessons as a kid and just wished you were playing your NES.

+ Go to makezine.com/14/diymusic_wiimote for Bill Byrne's OSCulator documents, helpful resources, demos, and the Painful Leg Injuries' recent album of Wiimote-controlled music, *The Rich Man's Godforsaken Driver's Seat*.

Bill Byrne is a multimedia artist, motion graphics designer, and educator who performs experimental electronic music with his wife, Suzanne, as the Painful Leg Injuries.

MOLECULAR GASTRONOMY

Spherify your food for a new culinary experience. By Michael Zbyszynski

There is a movement in the cooking world called Molecular Gastronomy. The term was coined by Nicholas Kurti and Hervé This, and it has become associated with chefs like Ferran Adrià at El Bulli in Spain, Heston Blumenthal at The Fat Duck in England, Wylie Dufresne at wd~50 in New York, and Homaru Cantu at Moto in Chicago.

Essentially, it involves applying scientific techniques to the cooking process. One of the more interesting techniques is the use of common substances to control the texture of foods, often in surprising ways.

You don't need a chemistry lab to pull off such effects. With a few inexpensive tools and chemicals, it's possible to use spherification to make all kinds of "caviar" (and other shapes) in your own kitchen.

In this project, I'll explain how to make a "spherification array" that allows you to quickly create many pieces of caviar. Next, I'll show you how to use it to make juice caviar and a molecular mojito.

Making a Spherification Array

Squirting the first dozen spheres out of a syringe is pretty fun. But it's somewhat time-consuming to make a large amount of caviar with this method and it's impossible to leave all the spheres in the setting bath for the same amount of time if it takes most of a minute to squirt them all out. That was unacceptable to me, so I wanted a different solution.

I found a product called the EZ Pipette 96 that is essentially a 96-headed syringe, made just for this purpose. I wasn't sure I wanted to spend $50, though, and then I saw a video on YouTube that showed chefs in Tokyo making carrot caviar using an array of syringes in an acrylic stand. That seemed perfect, and easy to build.

Photography by Henry Zbyszynski

MATERIALS

Mojito ingredients: limes (6), club soda, bundle of mint, granulated sugar, light rum, dark rum, water
Carrot juice for experimenting
Specialty food items: sodium alginate, calcium chloride, sodium citrate I found all these at a store in San Francisco called Le Sanctuaire (le-sanctuaire.com). I've also seen them sold at WillPowder (willpowder.net) and Texturas (texturaselbulli.com). Prices vary; I paid about $16 for sodium alginate and $5 for the others.
Acrylic basketball display case from TAP Plastics, tapplastics.com
9"×9" sheet of acrylic I had this lying around, but they're easy to get from TAP.
35cc syringes with catheter tips (16) from Small Parts, smallparts.com
Light box (optional)

TOOLS

Very accurate scale This kind of work requires precision to 1/10 of a gram, which is beyond what most baking scales offer. I got an Escali L600 laboratory scale, $60–$85, for a couple of reasons. It has the right precision and 600g capacity (high for such a precise scale), has some nice features like tare and counting, and looked more like a kitchen scale and less like a drug scale.
Big syringe with catheter tip, or squirt bottles Before I got the 35cc syringes for the spherification array, I picked up a 60cc syringe from Le Sanctuaire. I've also heard of people using traditional squirt bottles, like the ones used for ketchup and mustard. These are generally handy kitchen tools, but the syringes make me feel like a "real" chemist.
Kitchen tools: bowls, spoon, strainer, saucepan, immersion (stick) blender, measuring cup
Shop tools: Tape measure, calipers, combination square, sharp marker, drill, small pilot bit, 1" Forstner bit, chisel, clamps, scrap wood

1. Measure the syringes. I used a cheap set of digital calipers to find the diameter of my syringes: a little bigger than 1". I needed to drill holes in the acrylic case that would fit the syringe while letting its flange rest on the acrylic. I decided to use a 1" Forstner bit.

2. Mark the locations for the holes. I wanted a regular grid of syringes, so I marked the center of each edge of the case top with a small, permanent marker. I decided to make the centers of the holes 2" apart, so I marked 1" out from the center in either direction, and then 2" out from those marks. At this point I had 4 evenly spaced marks along each edge. I used a combination square to extend these marks to the 16 points where they meet. These would be the locations for my 16 syringe holes (Figure A).

3. Drill the holes. Drilling acrylic is somewhat of a pain; I actually tried 5 different drills. My crank-style hand drill did a great job of making pilot holes, but my brace couldn't really get a grip on the big 1" Forstner bit. I switched to a cordless drill, which would have been the perfect tool had the battery not been flat. The variable speed of the drill allowed me to go slowly, and not melt the plastic. After a second cordless drill ran out of juice, I switched to a corded drill. This made holes very quickly, but left a bunch of melted residue around the edges. I pulled off a lot of this by hand before it hardened, and carefully scraped off the rest with a chisel.

4. Install the syringes. The syringes go through the holes, pointy side down (Figure B). I found a scrap piece of acrylic to rest on the plungers of the syringes. Now my Spherification Array was ready to dramatically plunge 16 syringes. The drama is greatly enhanced if you do the plunging on top of a light box.

Spherification, or Making Caviar

One of the dramatic presentations from El Bulli is the creation of small, caviar-like spheres from fruit juices (Figure C). These spheres have a gelled exterior that pops when you bite it, releasing the flavor inside. The basic recipe is simple, but requires fairly careful measurement and timing. I did my first experiments with carrot juice, but the procedure below will work with any liquid, provided it's not acidic, doesn't contain calcium, and is water-based. There are ways to work with those more challenging ingredients, which I'll explain later.

1. Mix the juice. Measure out 250g of juice into a measuring cup on the scale. Gradually add 2g of sodium alginate. This is a very small amount, so be careful. Put that mixture in a bowl, and blend thoroughly with an immersion blender. A regular blender would also work, but it would get a lot more air into the mixture. Depending on the type of juice you use, you might want to strain this mixture after it's blended. Set the mixture aside to let some of the air escape.

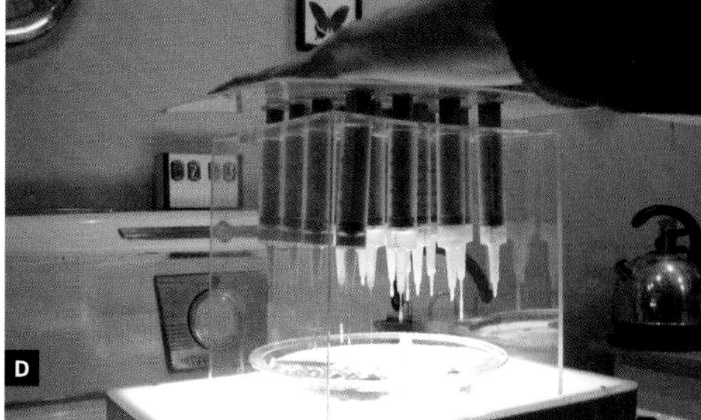

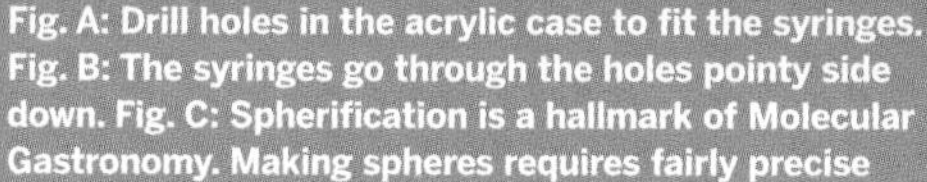

Fig. A: Drill holes in the acrylic case to fit the syringes. Fig. B: The syringes go through the holes pointy side down. Fig. C: Spherification is a hallmark of Molecular Gastronomy. Making spheres requires fairly precise amounts of chemicals, so a high-quality scale is a must. Fig. D: Squirt drops into the setting bath, controlling the size of the drops by the pressure you put on the plunger. Make the drops about the size of salmon eggs.

2. Mix the setting bath. Measure out 250g of room-temperature water into a cup on the scale. Dissolve 2.5g of calcium chloride into the water. Pour it into a bowl and add 250g more water, so the total is 500g of water.

3. Prepare a rinsing bath and wait. Fill another bowl about 2/3 full of water; this will be the rinsing bath. At this point I wait about 10 minutes to let everything dissolve, and to let air bubble out of the juice mixture.

4. Drop the juice into the setting bath. Suck up a bunch of carrot juice mixture into the syringe (or squirt bottle). Put a wide strainer in the bowl with the setting bath, so that it's mostly submerged. Slowly squirt drops of the carrot mixture into the strainer in the bowl, controlling the size of the drops by the amount of pressure you put on the plunger (Figure D). I try to make the drops about the size of salmon eggs.

5. Set. The calcium in the setting bath is now causing the alginate to gel. The ideal result is the thinnest skin possible that holds the shape of the drops but vanishes in a pop when eaten.

The thickness of the skin is controlled by the amount of time in the setting bath. For the carrot juice, I found that 30–45 seconds was perfect, but this time differs for other ingredients. A few minutes of setting yields a sphere that is gelled all the way to the center.

6. Rinse. Gently scoop the spheres out of the setting bath and into the rinsing bath. This should slow the gelling reaction enough for the skins to hold steady for a while. Eventually, the spheres will get all the way through, so they should be made immediately before serving.

7. Enjoy. My first product was a Carrot Martini, which I made by adding carrot caviar to gin and vermouth. Having succeeded with this, I wanted to figure out what else I could do. I was especially interested in working with citrus and alcoholic ingredients, which led me to the next cocktail, the mojito.

Mojito Moléculaire

The Mojito Moléculaire isolates the cocktail's into individual caviars. Each caviar has a distinct color and flavor, and the drink comes together on the tongue.

Photography by Michael Zbyszynski (A–C) and Adrian van Allen (D)

Molecular Mojito.

1. Mint caviar solution. The challenge of making mint caviar is to get a mint-tasting liquid. I tried infusing the rum, but wasn't very satisfied.

I decided to make a mint simple syrup. Simple syrup is a cocktail preparation where about a cup (or more) of sugar is added to a cup of water, and the mixture is boiled, then cooled. I added a whole bunch of chopped mint to the boiling water and strained it out after the syrup had cooled, leaving a sweet mint syrup that works in the spherification recipe.

2. Lime caviar solution. Limes are tricky because they're acidic. The trick is to add sodium citrate to the juice and alginate solution, which acts as an acid buffer.

After a number of attempts, I decided on 5g (or 1.6%) of sodium citrate to 250g of lime juice and 3g of sodium alginate. I started with about half that much citrate and alginate, and tested it. Initially, the spheres disintegrated when they hit the setting bath. I added more citrate and alginate, 0.5g at a time, and kept testing. I stopped when the spheres held together when dropped into the setting bath. I juiced 6 Persian limes.

3. Rum caviar solutions. I know that a mojito is traditionally made with light rum, but I felt compelled to use dark rum too. The 2 together give the drink more interesting color.

Whatever rum you decide to use, the problem is the same: sodium alginate doesn't like to dissolve in alcohol. I've seen one solution to this, called "reverse spherification," where the alginate is dissolved in the setting bath and another calcium salt (calcium gluconolactate) is dissolved in the liquor, which is probably thickened with xanthan. I'm curious to try that, but I went for a simpler procedure.

I dissolved 2g of sodium alginate in 125g of water, and then blended in 125g of rum with an immersion blender. That way, the alginate is dissolved before the alcohol is introduced. Since the rum in a mojito is heavily diluted, this is a reasonable compromise.

4. Make caviars with your Spherification Array. Making all this caviar would be tedious with just 1 syringe, but it goes quickly with the array. Place a pan full of setting bath beneath the syringes. I used a 9" pie plate, but a 9"×13" baking pan would be perfect. Fill each row of syringes with dark rum, light rum, lime, and mint/sugar solutions, respectively. Fill each syringe to the same level; I chose 20cc.

Place the sheet of acrylic on top of the plungers of the syringes. With both hands, slowly press down all the syringes using the sheet of acrylic, making sure that the solutions are dripping (not streaming) into the setting bath. Wait 60 seconds and skim out the caviar with a strainer. In one dramatic gesture, you will have made a round of drinks for the whole party.

5. Make the cocktails. Combine the mint/sugar, lime, light rum, and dark rum caviars in a glass with club soda and ice. (I made some water caviar, which I froze the night before, so the ice would match the rest of the drink.) Garnish with a sprig of fresh mint, and serve: Mojito Moléculaire.

To see the video of the Tokyo tapas bar that makes carrot caviar, go to youtube.com/watch?v=MskRv3mLxvg.

Michael Ferriell Zbyszynski (mikezed@mac.com) is a composer, sound artist, performer, and teacher in the field of contemporary electroacoustic music. Currently, he is assistant director of music composition and pedagogy at UC Berkeley's Center for New Music and Audio Technologies.

Photograph by Henry Zbyszynski

THE GOMICYCLE

A Honda Rebel 250 motorcycle goes electric. By Marque Cornblatt

While most civilians wait patiently for electric vehicles (EVs) to arrive at their local showroom, we makers take matters into our own hands. We've created a dizzying array of street-legal electric cars, scooters, and motorcycles, and there's even a National Electric Drag Racing Association.

I wanted to go electric for my own day-to-day transportation, but I didn't want to reinvent the wheel. So I researched the existing art, and purchased plans for the "El Chopper ET," a Honda Rebel 250 project that was developed by motorcycle EV conversion guru John Bidwell.

The plans turned out to be a little dated. They promised a quick and easy build, but the cost of the materials has risen, and some components are no longer available. I adapted and redesigned a few parts almost from scratch, but even though the project went "off the reservation," the El Chopper ET plans were a useful starting point.

Since I didn't have all the tools and skills needed to do the full conversion, I assembled an informal team. Gary Xaoui and Kat Townsend helped with fabrication, while Todd Kollin at Electric Motorsport in Oakland, Calif., where I bought most of the electrical components, contributed valuable advice.

Battery Matters

Projects like this always have tradeoffs like speed vs. range, sturdiness vs. weight, or quick acceleration off the line vs. high top speed. The biggest factor with this one was cost vs. performance, particularly regarding the batteries. In theory, a lithium-ion battery bike would rock, but the batteries alone would cost something like $10,000.

Photography by Marque Cornblatt and Jeff Kramer

MATERIALS

- **Honda Rebel 250 rolling chassis** We found a bike with a blown engine (Figure A) for less than $500 on Craigslist, craigslist.org.
- **12V, 50Ah sealed lead-acid golf cart batteries (4)**
- **Perm PMG-132 electric motor**
- **36V–72V PWM (pulse-width modulation) controller**
- **0KΩ–5KΩ twist-grip electric throttle (potentiometer)**
- **48V AC charger**
- **48V–12V DC-DC down converter, or additional small 12V battery** for lights, signal, and horn
- **½" steel angle beam, 80" long**
- **Thick, long zip ties**
- **¼" plate steel**
- **#4 welding cable and lugs**
- **Front/drive sprocket and rear/driven sprocket** The optimal number of teeth for each of these will depend on the specs of your motor, the size of your rear wheel, and your target top speed. Consult a gear ratio calculator, such as makezine.com/go/gearratio. You may need to order custom sprockets.
- ***El Chopper ET Builder's Guide*** by John Bidwell available from 21 Wheels, 21wheels.com
- **Honda Rebel 250 original factory service manual** Without this, the project would have been lost.

TOOLS

- Welder
- 4" angle grinder with cutting and grinding wheels
- Drill press
- Metal band saw
- Sawzall or other reciprocating saw
- Wrenches and other standard auto shop tools
- Heavy duty wire cutters/crimping tool
- Rags and solvent for cleaning
- Friends who can help

Fig. A: A used Honda Rebel 250 with a blown engine, from Craigslist.

The El Chopper ET plans estimated the cost for the whole project at $1,200.

The first version of my GomiCycle used a set of 4 big 80Ah lead-acid batteries, 2 of which were free; I already had them lying around from an old robot project. But they had suffered through the heat and corrosive dust of Burning Man and were in sad shape. We welded heavy, ultrasturdy trays for the batteries using 2" angle steel (Figure B), and held them down with strap steel and a padlock. Two batteries fit in the engine compartment and 2 more straddled the back, mounted like saddlebags.

But the batteries were a mismatched set, and their performance proved disappointing. For the next version, we bought new, smaller 50Ah batteries and made trays to hold them that weighed 80% less. The smaller battery set was less powerful, but the changes saved so much weight that the GomiCycle got roughly the same top speed and range as before, 40mph and 15–20 miles.

Mod the Frame

The first step in the GomiCycle conversion was to strip the frame (Figure C). This was surprisingly fun and fast to do with basic hand tools. Make sure you save all the nuts, bolts, and other bits and pieces for reassembly later. Clean the frame using rags and solvent. Then use the Sawzall and angle grinder to cut off the motor's mounting points and tabs.

The original plans called for chopping and lengthening the frame to fit 4 batteries in the engine compartment. Instead, we opted to retain the Rebel's original geometry. The relatively compact 50Ah batteries all fit in the compartment, arranged in 2 pairs. We made trays for them out of ½" steel angle and held them down with enormous zip ties.

We modified the frame's swing arm, the fork-like part that's mounted on shocks and holds the rear wheel. Following the original plans, we cut a rectangular hole in the arm for the electric motor to fit into, just in front of the rear wheel (Figure D). Then we shaped and welded a custom bracket out of steel plate to mount the motor (Figure E), making sure its sprocket would align with the rear wheel sprocket, which is critical for keeping the chain in line and at the correct tension during travel over rough roads.

Assemble the Power Train

The next step is to replace the motorcycle's entire original drivetrain — the engine, clutch, transmission, carburetor, and exhaust system — with 4 batteries,

Fig. B: The original, overbuilt battery tray for the engine compartment. Fig. C: The stripped motorcycle frame. Fig. D: The swing arm, with a clearance hole and mounting bracket for the motor. Fig. E: The electric motor, bolted to the mounting bracket on the swing arm. Fig. F: The GomiCycle's wiring, on an earlier version of the bike, with 80Ah batteries. We stuffed the electrical components into a camping cooking pot.

an electric charger and controller, and a motor with just 1 moving part, the hub. No oil is needed, and virtually no noise or heat is produced during operation. Clearly, this is not your father's motorcycle.

An unmodified Rebel has a small 12-volt battery under the saddle that powers the lights, horn, and starter. Our new electric drivetrain needs to operate along with this original electrical system. One option is to keep the systems separate, by simply adding a small 12V battery dedicated to the lights and such. This option simplifies the wiring, but also requires a second battery charger and more daily maintenance — a solution that I found inelegant.

Instead, we spliced a 48V-to-12V DC converter into the main power system and pulled our 12 volts from there. This taxes the batteries slightly, especially when the headlight is on, but everything recharges from a single standard wall plug. Xaoui figured out the wiring, which was way outside my comfort zone, and it took a few rounds of trial and error to get it right (Figure F).

Conclusion

The GomiCycle project was a challenge. It required a lot of people, skills, and tools. The original plans said that the bike would cost around $1,200, go 60mph, and have a 40-mile range, but ours cost almost $3,000, tops out at 40mph, and has a range of 15–20 miles. But the story is far from negative.

The GomiCycle is a great short-hop urban commuter bike with lots of torque and pep — it climbs San Francisco hills without difficulty, and people love it wherever it goes. It costs pennies per mile to ride, and its lack of emissions makes the wind smell that much sweeter.

WHAT IS GOMI?

Gomi is originally a Japanese word for dust or garbage, but it's now slang used to describe anything that people discard or no longer value. The word was introduced to English speakers by best-selling author William Gibson, who uses it to describe a near-future dystopia of our material culture gone haywire.

In my working, gomi is DIY. Gomi is tech-positive. Gomi breaks the rules, but is responsible. Gomi is an aesthetic choice — the best defense against mindless materialism.

—Marque Cornblatt

Marque Cornblatt is a conceptual artist, roboticist, and maker. He is the creator and host of *Gomi Style*, a DIY lifestyle and design video series (gomistyle.com).

ART WORK

Life Models

By Douglas Repetto

ARTISTS' STUDIOS HAVE TRADITIONALLY been full of reference materials like botanical drawings, medical texts, photographs, catalogs, and images clipped from magazines. Artists use these images as direct models for realistic renderings, but they often provide indirect inspiration as well; patterns in a botanical drawing might end up as abstract gestures in a painting, or shapes from a tool catalog might inspire sculptural forms.

Just as often, reference materials simply set the mood or tone in the studio; being surrounded by meaningful materials is an inspiration in itself. I recently visited the reconstruction of Francis Bacon's home studio at the Hugh Lane Gallery in Dublin, and was overwhelmed by the density and intensity of the materials (books, magazines, clothing, painting supplies, canvases) jumbled across every surface in the room. Although Bacon didn't often use images from these sources directly in his paintings, he was certainly inspired by having them around. He said: "I feel at home here in this chaos because chaos suggests images to me." Interestingly, the rest of his small home was rather tidy and uncluttered; the chaos of his studio seemed a conscious technique, a key part of his process as a painter.

In addition to a studio littered with reference debris, many artists now also look to the seemingly unlimited resources of the internet: image searches, video sharing sites, scientific and historical databases, and so on. Their studios might still be chaotic, but now their computer desktops are as well.

Ceramics artist Christopher Russell has been working on a series of hyperdetailed, incredibly lifelike (especially given the fact that they're monochromatic!) bee sculptures, including intricate sections of beehives, bees on flowers, their little legs laden with pollen, and large blowups of grains of pollen. He mostly uses traditional, labor-intensive manual techniques to sculpt the pieces, often working from photographs or illustrations in books.

But recently he's also been using images from online science resources, like the PalDat palynological (the study of pollen and spores) database, which is where he found high-resolution electron microscope images that he used to create the giant pollen grains. The site has a graphical search feature that made it easy for a nonexpert to navigate the huge variety of images by searching for the sorts of shapes and textures he was interested in.

Russell told me, "One of the reasons I enjoy working from life is that nature comes up with textures and forms that I just wouldn't ever imagine myself, or think I could produce myself. When I saw the pollen images, my first reaction was that they would be impossible to reproduce in clay. Then my mind began to figure out production techniques, over time, while I was in the shower or walking, or driving. At some point I thought, 'I should give it a try,' and ten days later I had a ball covered in spines."

Russell is using traditional techniques to create objects based on very contemporary sources — electron microscopes have only been around for 70 years or so, and the PalDat website went live in 2005. Before electron microscopes, we didn't really know in detail what pollen grains looked like, and before the internet, getting access to scientific images was often an arduous and expensive proposition. PalDat is free for noncommercial use (although many scientific resources, particularly peer-reviewed journals, are still notoriously expensive).

Caitlin Berrigan's recent *Viral Confections* project takes a similar approach to gathering source data, but her method of translating 3D data to a physical object uses more contemporary techniques. While reading about similarities between the structure of some viruses and Buckminster Fuller's geodesic domes, she started imagining viruses as structures or shelters at a human scale, turning the relationship between virus and human inside out. Diagnosed with hepatitis C herself, she decided to make an artwork that addresses both the formal aspects of the virus and its often devastating consequences in people. She looked online for information about the structure of the virus, but found little about it.

After consulting with a biologist, Berrigan decided to use the dengue virus, a close cousin of hepatitis C, instead. She was able to get structural information about dengue from the online RSBC Protein Data Bank, but it was in a format that wasn't easily imported into the commercial 3D applications she used. Finally, after a convoluted series of translations and interpretations, she ended up with a 3D model of the virus that she could send to a rapid prototyping machine.

She printed out a large-scale replica, and then used traditional mold-making techniques to create food-grade silicone molds, which she used to cast edible chocolate copies of the virus. She now hosts tea parties where guests are invited to "befriend the virus" by consuming the chocolate truffles and talking with the artist about the issues surrounding hepatitis C.

The nature of studio hoarding changes when all the information you need is "out there" for the taking, 24/7. That one iconic image on your wall takes on a different meaning when it's surrounded by thumbnails of other images that a search engine has decided are similar.

It's a bit startling to realize that the first glimpse of the New World that most Europeans got was via landscapes painted in the Americas and shipped back to Europe with round trip times of months or even years. Now, images from Mars arrive daily, and an artist who wants a better look at the rings of Saturn need only point her browser at the NASA website. What does access to all that data, all those images and sounds and 3D models, not to mention live webcams, tactile feedback devices, and perhaps soon, online smell-o-vision, mean for artists? There's only one way to find out.

RESOURCES

Christopher Russell: russellproject.com
PalDat palynological database: www.paldat.org
Caitlin Berrigan: membrana.us
Research Collaboratory for Structural Bioinformatics (RCSB) Protein Databank: pdb.org

Douglas Irving Repetto is an artist and teacher involved in a number of art/community groups including Dorkbot, ArtBots, Organizm, and Music-dsp.

Photography courtesy of Caitlin Berrigan and Christopher Russell

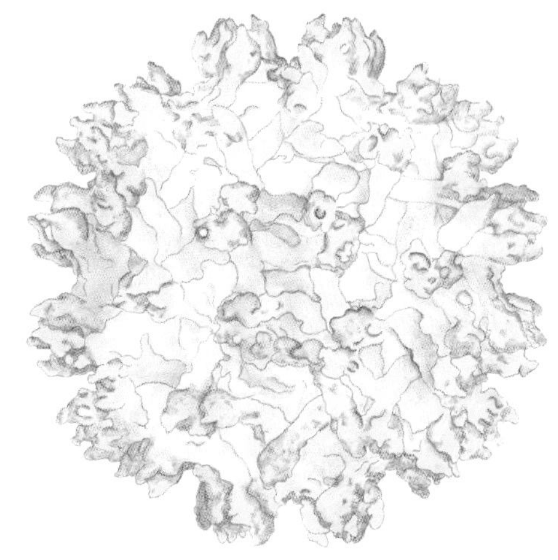

TOP: Caitlin Berrigan's 3D virus model, which she used to make the molds for her edible chocolate hepatitis C viruses, pictured here (middle) in their original packaging. BOTTOM: Christopher Russell's glazed white earthenware pollen grains are about 4" in diameter. They're sculpted by hand using traditional ceramics techniques.

BY TOM OWAD

Three-Dimensional Printing Methods

3D printers print layer on top of layer, slowly building a three-dimensional object. A plethora of materials and methods are used to build these layers.

The Stratasys Dimension is a commercial 3D printer that uses ABS plastic. The ABS filament comes in a self-loading cartridge, and is fed into a heater block by two drive wheels. In the heater block, the ABS is heated to a semiliquid state and extruded through the tip, with layers as thin as 0.01".

The open-source RepRap (reprap.org) project uses similar technology. The filament is a polymer, 3mm in diameter. This is fed into a chamber, held tightly against a drive screw. As the drive screw turns, it pushes the filament down into a heated barrel. The heat comes from a strip of nichrome wire that is wrapped around the barrel. At the tip is a nozzle with a diameter typically between 0.25mm and 1mm (smaller is more precise, but takes more time). A thermistor is embedded in the nozzle, so it's possible to adjust the temperature for different polymers.

One of the easiest polymers to work with is Polymorph, which is marketed as Friendly Plastic or ShapeLock in the United States. Polymorph extrudes easily, is relatively strong, and melts at just 140°F. Unfortunately, it's also expensive and deforms if regularly exposed to temperatures over 100°F.

RepRap enthusiasts have also experimented with HDPE, ABS, and PLA (polylactic acid). PLA is the most interesting, as it's biodegradable and you can make it yourself. PLA is formed by heating lactic acid with stannous chloride; and lactic acid can be created by fermenting milk or starch. The process is sufficiently complex that it might not be terribly practical in small quantities, but it opens up some interesting possibilities for local production.

Another material the RepRap project has been experimenting with is EcoComp UV-curable resin, from Sustainable Composites Ltd. The resin is mixed with glass filler to make a paste, and hardens when exposed to ultraviolet light. A ring of UV LEDs mounted around the nozzle hardens the paste as it's extruded. EcoComp resin is made from plant oil, and like PLA, will biodegrade if composted, making the material carbon-neutral. EcoComp resin, however, is very stable underground. If buried, it would actually *reduce* the amount of carbon in the atmosphere.

The Fab@Home (fabathome.org) project uses a much simpler extrusion method than the RepRap: the mechanism is simply a motor-controlled syringe. A set of 30 syringe barrels and pistons costs $36, so it's easy to experiment with different materials.

This can lend itself to a lot of fun — peanut butter, icing, toothpaste — if you can squirt it out of a syringe, it'll extrude. Making something sturdy out of peanut butter might be difficult, but the fun is in the experimentation. Betty Crocker Easy Squeeze Decorating Icing works well and could be great for 3D decorations.

> Build one platform, and you can experiment with everything from cookie dough to laser sintering.

Silicone also works well in the Fab@Home. It cures within 24 hours to a somewhat rubbery material. It's also possible to get conductive silicone, which enables circuits to be embedded. For making solid models, FabEpoxy, from Kraftmark, looks like the most promising material. It's a two-part epoxy that is very durable and machinable, and can be painted when cured.

All of these materials can be extruded at room temperature, which significantly reduces their complexity. Ironically, the Fab@Home syringe tool is very expensive to build, because of the $130 linear motor that plunges the piston. That's a lot of

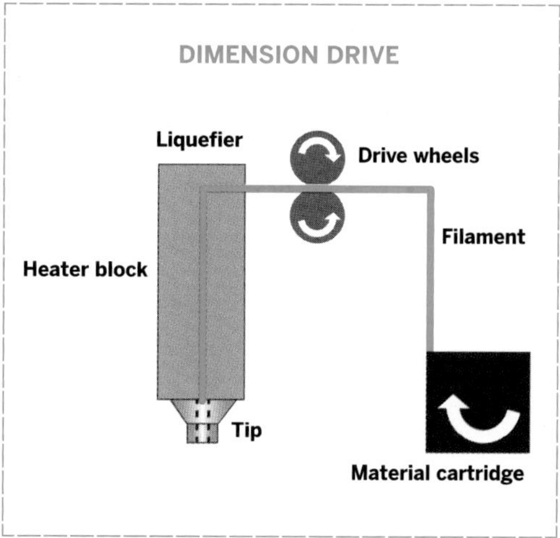

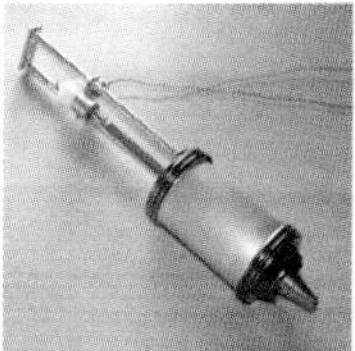

CLOCKWISE FROM LEFT: The Fab@Home extruder is a syringe driven by a linear actuator; the Stratasys Dimension 3D printer extrudes a filament of heated plastic; the author's DIY cookie press, retrofitted for computer control with a $6 motor.

money if you want to have numerous interchangeable extrusion heads.

Fortunately, there are lots of ways to hack around this. The first method that came to my mind is to adapt an old cookie press. These sell for $5 or $10 on eBay, and all you have to do is drill and tap a few holes to mount a motor. I chose a $6 Solarbotics GM3. My cookie extruder is pretty large, but a similar method could be used with a captured nut and a screw, on the standard Fab@Home syringe.

This method of fabricating that I've been describing is called fused deposition modeling (FDM). A similar method, called precision droplet-based net-form manufacturing (PDM) is being developed for use with metals by Melissa Orme, Qingbin Liu, and Robert Smith at the University of California, Irvine.

Their work has been with aluminum, but the process is expected to be suitable for a wide variety of metals. The molten aluminum is held in a graphite-lined canister with a vibrating plunger rod at its center. Perturbations from the plunger rod cause uniform droplets of aluminum to break away and exit the canister through a small hole in the bottom. In other words, it squirts out beads of molten aluminum. These beads cool quickly as they exit, and can be used to build objects in much the same manner as the plastic. This method is still being refined, but there are other additive methods in commercial use, such as selective laser sintering (SLS).

In the SLS process, a layer of powder is laid upon the bed. A laser then traces the shape of the desired object over the powder, fusing it together. Additional layers are laid down and then fused, building the object. This process can be performed with numerous materials, including plastics and metals.

A very similar technique is used in Arcam's EBM (electron beam melting) machines, where an electron beam operating in a vacuum fuses titanium or chrome powder. While lasers and electron beams pose a relatively high barrier to entry, the same concepts can be applied to a simple heat gun. The CandyFab Project (candyfab.org) uses a $10, 500W air heating element to fuse together models made of pure sugar, with excellent results (*see MAKE, Volume 12, page 38*).

All of these 3D printing techniques share an important characteristic: they require a 3-axis Cartesian system to position the tool, whether that tool is a syringe or a focusing lens. Build one platform, and you can experiment with everything from cookie dough to laser sintering.

Tom Owad is the owner of Schnitz Technology, a Macintosh consultancy in York, Pa. He spends his days tinkering and learning, and is the owner and webmaster of applefritter.com.

Photography and illustration by Tom Owad

PRIMER
FLEX ware
FLEX ware

Solar Power System Design

How to use solar panels to supplement your home or workshop electricity needs.

By Parker Jardine

In the first part of my solar power Primer, I showed how to make inexpensive photovoltaic (PV) panels (*see MAKE, Volume 12, page 158, "20-Watt Solar Panel"*).

Here, I'll explain how I incorporated them into a complete solar PV power system. While this article provides installation tips and general how-to information, it's not a step-by-step guide to building the complete system. Rather, it's an introduction to a complex project that could easily fill a book.

»

WARNING: I'm not a certified solar PV installer or certified electrician, nor do I know every detail pertaining to electrical codes. If you decide to make your own PV system, take the proper safety precautions and adhere to the electrical codes for your area. Failure to do so could result in serious injury or death by electrocution.

Photography by Gunnar Conrad

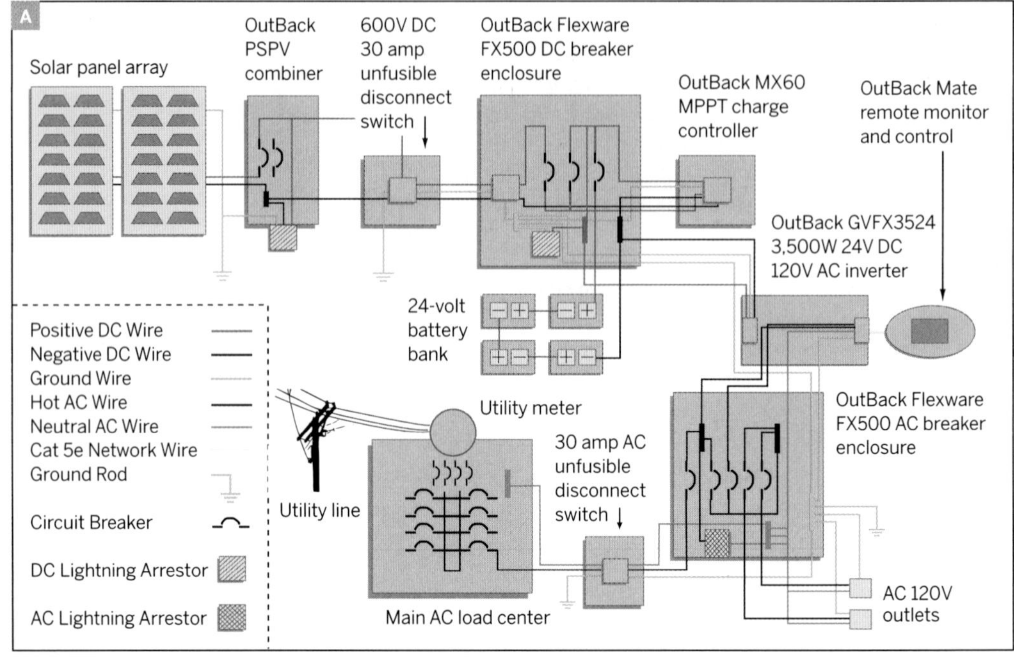

PLAN YOUR POWER NEEDS

First, identify your overall goal, obstacles, equipment and hardware needs, equipment placement, and solar panel array location.

There are 3 basic types of solar PV systems: off-grid solar PV systems, grid-tied systems with no battery backup, and grid-tied systems with battery backup (which is the type I built).

Build It Solar (builditsolar.com) is a great resource for all 3 types of solar projects.

MATERIALS

My system mainly consists of OutBack Power Systems equipment (outbackpower.com), including the inverter, breaker boxes, charge controller, and other components. You can buy these components at many different locations. A few online stores I recommend are Affordable Solar (affordable-solar.com), the Alternative Energy Store (altenergystore.com), and The Solar Biz (thesolar.biz).

1. SOLAR PANEL ARRAY

Solar PV systems normally consist of multiple solar panels connected together in series and/or parallel to form a solar PV array (Figure A). A solar charge controller protects the batteries by regulating the current and voltage coming from the solar array; its specifications usually determine how the solar panels must be configured. For instance, some charge controllers only accept specific DC voltage inputs of 12V or 24V. Therefore, you can connect two 12V solar panels in series to create a 24V solar PV string. You can then combine multiple 24V strings in parallel and connect them to a PV combiner enclosure (which takes the input from multiple solar panels and combines them into one DC output).

Note that 24V is too low for certain system designs, because when you increase the current on a wire with only 24V, you decrease efficiency and increase voltage loss. A larger-gauge wire would offset power loss by reducing resistance, but the best solution is to use a charge controller that accepts a dynamic DC voltage and steps down the voltage to charge the battery bank. Dynamic charge controllers like the OutBack MX60 enable you to stack solar panels in series up to 150 VOC (open circuit voltage). This configuration allows multiple solar panel stacking options and saves money by reducing the wire size needed.

2. PV COMBINER ENCLOSURE

The OutBack PSPV combiner box (Figures B and C) combines multiple strings of solar panels together in parallel to form a solar array. Each string of input wire will have a specific voltage and amperage, depending on the solar array configuration. While it's OK to have different amperage ratings on the input strings (because the amps get added after leaving the combiner box), you should not combine multiple strings of solar panels with different voltages. I have two 10-amp breakers in my combiner box for my 2 strings of solar panels.

The gray cylinder (circled in Figure C) is the DC lightning arrestor. Its 3 wires go to the positive bus, negative bus, and ground. It protects your equipment from lightning damage and is required by code.

3. GROUNDING EQUIPMENT

Proper equipment grounding is a must. It reduces and prevents shock, trips a breaker if a ground fault occurs, and limits the potential for equipment damage by lightning. Below the combiner box, I've driven an 8' copper ground rod into the ground. Underground, I connected a direct burial ground lug to a 6-gauge copper wire connected to the combiner box at the designated location. I screwed code-compliant ground lugs into each aluminum rail that touches a solar panel. A copper wire extends from these solar panel ground lugs to the ground rod below the combiner box. Read John Wiles' article, "To Ground or Not to Ground: That Is Not the Question (in the USA)," available at makezine.com/go/wiles.

4. DC DISCONNECT SWITCH

How do I ground my power center inside my shop, route DC cable into my workshop, and easily disconnect the DC power source for safety? With Square D's HU361RB 600V DC 30A unfusible disconnect switch (Figure D). The cable entering the box from the bottom right is the solar input from the OutBack PSPV combiner. These wires connect to one side of the disconnect junction and to the ground bus. Three additional wires connect to the other side of the disconnect junction and the ground bus, then run into the building through the conduit. Check with your local inspector to see if PVC conduit on the DC side is allowed to enter the building.

Note the bare copper wire (Figure E) that's connected to the ground bus bar, and then runs down the side of the building to connect to another 8' ground rod. The NEC requires a separate ground rod for the AC side of the system, and these 2 grounds need to be bonded together. Again, check with your local inspector for details.

B

C

D

Photography by Parker Jardine

E

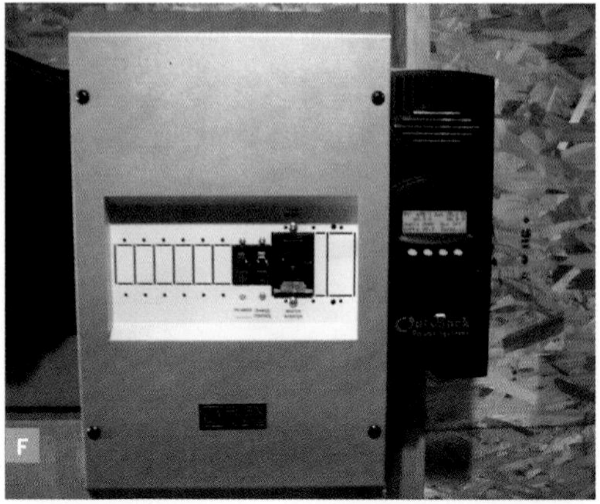
F

G

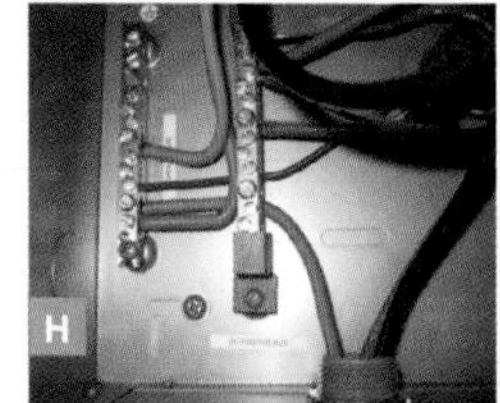
H

5. DC BREAKER AND CHARGE CONTROLLER

The Flexware FX500 DC enclosure (Figures F, G, and H) serves multiple purposes with room for expansion. The 3 wires that enter the building from the DC disconnect switch enter into this enclosure. The green wire connects to the ground bus bar. The black wire connects directly to the negative terminal of the solar charge controller. The red wire connects to one side of a 60A breaker. The other side of the breaker then connects to the positive terminal of the charge controller (Figure F, black enclosure on the right). This breaker allows you to protect your solar charge controller, and disconnect the solar input when maintenance is required.

The sun's input to your panels varies throughout the day — the sun will shine on the panels at one moment, and be covered by clouds the next. This up-and-down cycle would shorten a battery bank's lifetime without a solar charge controller, which regulates the charge to the battery bank.

The controller has 2 terminals that connect to the battery bank. The black wire from the charge controller's negative terminal connects directly to the negative bus in the DC enclosure. The red positive wire on the charge controller connects to one side of a 60A breaker in the DC enclosure. (The other side of the breaker connects to the positive bus bar in the enclosure.) This additional breaker protects the charge controller on the battery side, and allows for charge controller isolation during maintenance.

The enclosure's negative bus bar is connected to a DC current shunt included with the FX500 DC enclosure. The shunt measures current and voltage. When used in conjunction with a battery meter, it can provide valuable information relating to your battery bank. The shunt connects to the negative end of the bank and to the negative side of the inverter.

Similar to the PSPV combiner box, I have another DC lightning arrestor in my FX500 DC enclosure box. You can never be too careful when it comes to protecting your expensive equipment. Connect the red wire of the lightning arrestor to the positive bus bar, black wire to the negative bus bar, and green wire to the ground bus bar.

If you have a roof-mounted solar array, then the NEC requires a Ground Fault Protection (GFP) device. This device disconnects the solar array when a ground fault is detected. For details on this and other grid-tie subjects, see John Wiles' "Making the Utility Connection" at makezine.com/go/wiles2.

6. BATTERY BANK

My battery bank consists of four 6V Trojan T125 lead-acid batteries wired in series (Figure I). Each

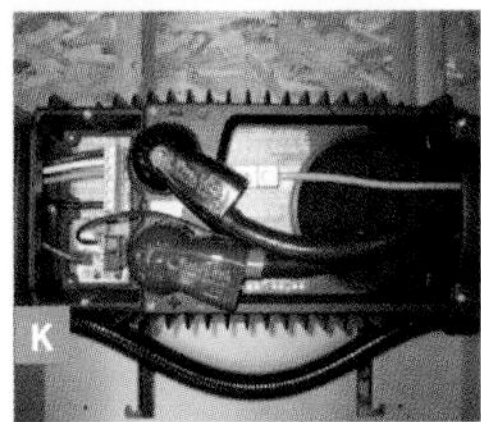

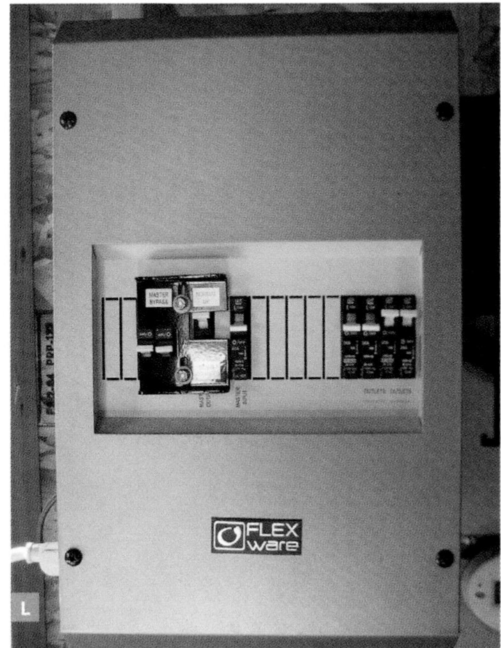

battery has a 240 amp-hour rating at 6V, so the 4 batteries connected in series yield a power rating of 24V at 240Ah. This particular lead-acid battery is considered a "deep cycle" battery, which is different than a car battery because you can drain its energy capacity by up to 50% without damaging the battery.

The Trojan T125 battery requires adding distilled water on a regular basis, because charging and discharging consumes water and releases small amounts of hydrogen into the air. Since hydrogen is flammable, venting is required by code if you have a battery bank inside a house or workshop. I recommend reading "Battery Enclosures" from *Home Power* magazine #119 for details on how to build your own code-compliant battery enclosure.

Using three 10" 4/0 gauge cables, I connected the positive terminals to the negative terminals of all 4 batteries. This series connection leaves open 1 positive and 1 negative terminal, to which the 72" cables coming from the DC enclosure box connect.

7. INVERTER

The components in a solar system must be carefully matched with the type of inverter you use. My inverter requires a 24V battery source. Many other items, such as inverter cable gauge, breaker ratings, solar charge controller voltage, and the solar panel configuration, are all tied to the inverter.

One of the main features of the OutBack GVFX3524 inverter (Figures J and K) is that it can charge batteries as well as sell power back to the utility grid. I'm into electronics, and being able to utilize DC power for my electronics projects in addition to selling excess power back to the grid was my ultimate goal, which is why I chose the GVFX3524.

Most grid-tie inverters don't work with battery banks, which means they're usually more efficient than off-grid inverters, because 10% of the electricity is lost when charging a battery (and further losses occur when inverting DC power to AC). However, grid-tie inverters that don't use batteries require a DC input voltage as high as 150V to 250V. This high voltage requires a lot of solar panels wired in series to meet the minimum voltage. Because I didn't want to buy or build nine 24V solar panels, I decided to use the OutBack inverter instead of a typical grid-tie inverter.

8. AC BREAKER ENCLOSURE

An OutBack AC enclosure (Figure L) connects the inverter to various AC loads in my workshop. Any standard AC breaker box would work, but the OutBack AC enclosure has punch-out holes that line up

M

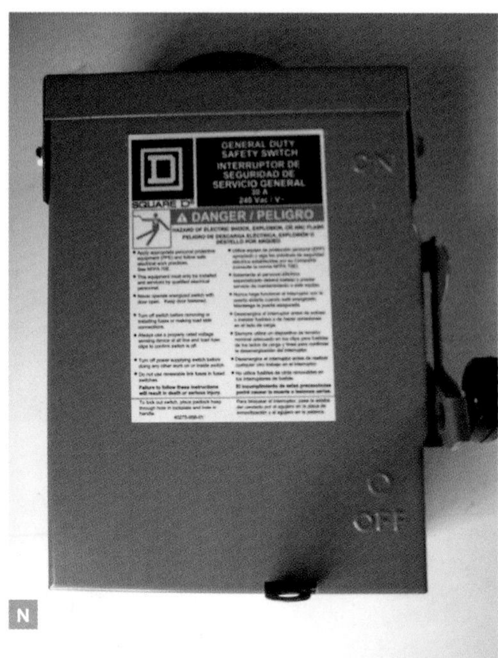

N

perfectly with the inverter, so no wires are exposed between the 2 units.

OutBack makes an input/output bypass (IOB) assembly kit designed to work with the AC breaker enclosure, specifically for grid-tie scenarios. A bypass assembly kit is a set of breakers that can be used to isolate your inverter from the grid when maintenance is required. The AC loads that the inverter normally powers through separate breakers get switched to being powered by the grid. You can see the black bypass plate and set of breakers in Figure L.

9. GRID CONNECTION

One of the most difficult parts of this project is configuring the solar PV system to sell power back to the grid. It can be beyond the realm of some solar installers and electricians alike. You must contact your electrical utility company to obtain permission to connect to the grid, and figure out what needs to be done to receive any available financial incentives, rebates, and tax credits. I also recommend showing your plans to a licensed electrical inspector, who can tell you what you need to do to pass inspection.

Following are a few things I learned while working with the local electrical inspectors. When digging a trench from one building to another, the trench must be 2' deep. Grey PVC pipe can be Schedule 40 underground but must be Schedule 80 above ground. Conduit expansion joints must be used with trenched conduit in cold climates above ground level, to prevent the conduit from cracking from temperature changes. All AC cable runs are required to have an AC disconnect at the point of entry to any building. This disconnect can be inside or outside of the building.

I'm using another small load center to meet the National Electrical Code's point-of-entry disconnect requirement. You can see the black wire (Figure M), which runs to the AC In bus bar in the Flexware AC500 box.

Connecting to the utility grid has been one of the most confusing parts of the whole process. The NEC 2005 code has specific requirements that regulate the amperage load on the load center's bus bar and conductors. See makezine.com/go/nec for details on the NEC 2005 Residential Utility Connection options and requirements. Again, see "Making the Utility Connection" at makezine.com/go/wiles2.

10. AC DISCONNECT

My local electric utility requires an AC disconnect switch (Figure N) on the outside of my workshop or house before connecting to the grid. You can use Square D's DU221RB 30A 120/240V AC 2-pole disconnect switch on the outside of your house or shop before connecting to the main AC load panel.

11. OPTIONAL ACCESSORIES

Other items I'm using in my power center are the Trimetric 2020 battery meter (Figure O), and the OutBack Mate remote monitor/control and temperature sensor (Figure P). The Trimetric 2020 is primarily for monitoring battery conditions including voltage, charge percentage, amps in or out of the battery bank, and total amp-hours used; it connects to the DC shunt inside the Flexware FX500 DC enclosure box. The OutBack Mate is used to control and program the OutBack inverter, via a standard Cat 5e network cable.

12. WIRE INFORMATION

The type and size of wire that you use throughout your solar installation is very important. An excellent wire efficiency/voltage drop spreadsheet for copper wire is available at makezine.com/go/homepower.

I'm using 6 AWG THHN wire throughout the DC and AC enclosures. I'm using 6 AWG THWN wire in wet locations, where conduit is used outside. The ground wire should be either 6 or 4 AWG bare or stranded copper. Throughout the inside of the workshop, I'm using Romex 12/2 for 20A AC circuits, and 14/2 for 15A AC circuits. The battery and inverter cables are 4/0 AWG copper wire.

13. OFF-GRID VS. GRID-TIE

If you live in a rural area outside the range of the utility grid, the electric company could charge you $20,000 or more to run an electrical line to your property. So you could either spend the $20,000 to connect to the grid and pay a monthly electricity bill, or invest it in clean solar PV technology and reduce or eliminate your monthly bill.

However, an off-grid home requires a substantial (and therefore costly) battery bank with enough capacity to power the home for extended days without sunshine.

A grid-tie system is cheaper because you don't need to buy and maintain a large set of expensive batteries. When sending power back to the grid, you will offset your current electrical usage, and if your solar electricity production surpasses your usage, your utility meter will spin backward. At this point you'll be selling power back to the utility company.

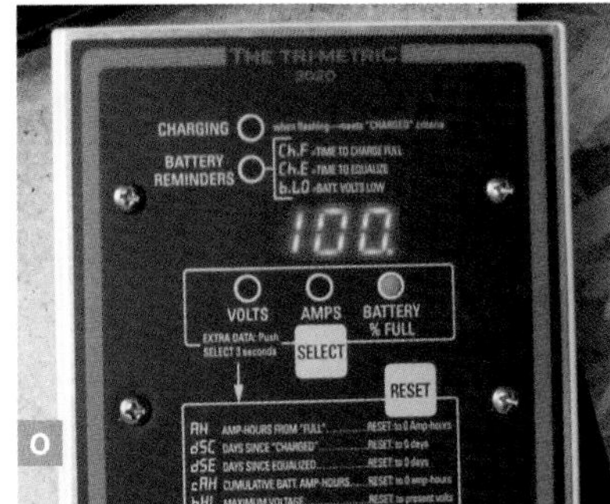

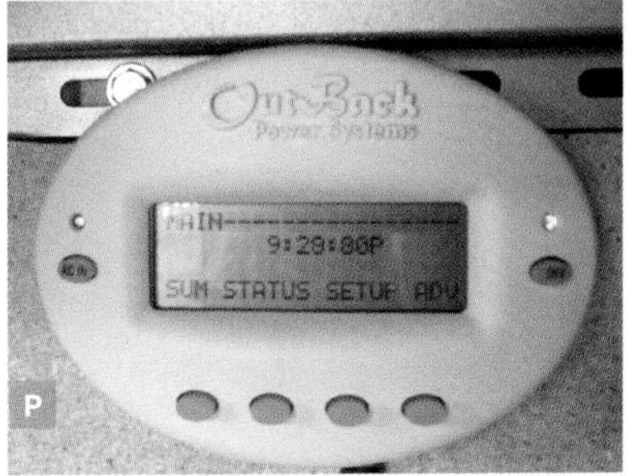

RESOURCES

Photovoltaic Systems and the 2005 National Electric Code (NEC) makezine.com/go/pvnec

Colorado Residential Wiring Guide makezine.com/go/colo

Home Power Magazine homepower.com

John Wiles' "To Ground or Not to Ground" makezine.com/go/wiles

Conduit Calculator makezine.com/go/conduitcalc

Electric Service Information makezine.com/go/elecservice

Adventures of Conduit Phil makezine.com/go/conduitphil

Parker Jardine is a systems administrator in Durango, Colo. He enjoys cycling, kayaking, rock climbing, electronics, and renewable energy. He would like to thank Roger Dodd, solar PV expert extraordinaire, for his help over the years.

CUT OUT AND TAPE ON
CUT OUT AND TAPE ON TOP OF CD

HOWTOONS.COM
il•lu•sion
[ILL-OO-ZHUHN] – NOUN
a perception, as of visual stimuli (optical illusion), that represents what is perceived in a way different from the way it is in reality.
–DICTIONARY.COM
GIVE IT A SPIN
D.I.Y.
MATERIALS
PUSH-PIN
TAPE
CD
STICKY SIDE

By Lee D. Zlotoff

Under Siege

The Scenario: After an exhausting day of work — that included a tiff with your steady romance of the last few months — you drag yourself home, determined to put the day behind you with a few drinks and a couple hours of mindless time parked in front of the tube. You thought about stopping at the gym, but the car just wouldn't go there.

So you plug your cellphone in to recharge in the kitchen, grab some munchies and the requisite inebriants, and assume the position on the living room sofa to slouch toward nirvana. And you're just about there — or at least on the edge of dozing off — when there's a knock on your door.

Thinking it might be your paramour coming by to kiss and make up, you pull yourself vertical to open the door when ... you're suddenly knocked back into the room by a pair of burly guys who barrel their way in, shouting things at you that you'd rather not hear.

Scrambling to your feet, you make a break for the back door, only to see a third guy waiting outside to cut off your escape. Your only option is to duck inside the laundry room, slam the door, and shove the washer against it with the strength of the adrenaline pump you have going. Fortunately for you, the door opens inward and the hinges are on your side. Unfortunately for you, it's a windowless room with only one way in and out.

The Challenge: Your newfound friends are on the other side of that door, working hard to get in since they seem to know you have a heavy-duty wall safe in your office that they want to "talk" to you about. The washer may not hold them off for long, and with no landline in here and your cellphone in the kitchen, 911 is not an option — nor is screaming for help, since the room is well insulated and your neighbors are just beyond earshot. Which means you'll have to find another way to signal for help, or a way to deter these guys long enough that they'll settle for hauling off what they can carry, and leave you in one piece. Any ideas?

What You Have: In addition to what was already mentioned, you have whatever else one might typically find in a modern laundry room — including a working sink. So take a deep breath, think carefully, and make it work!

Send a detailed description of your MakeShift solution with sketches and/or photos to makeshift@makezine.com by Aug. 29, 2008. If duplicate solutions are submitted, the winner will be determined by the quality of the explanation and presentation. The most plausible and most creative solutions will each win a MAKE sweatshirt. Think positive and include your shirt size and contact information with your solution. Good luck! For readers' solutions to previous MakeShift challenges, visit makezine.com/makeshift.

Lee D. Zlotoff is a writer/producer/director among whose numerous credits is creator of *MacGyver*. He is also president of Custom Image Concepts (customimageconcepts.com).

Photograph by Jen Siska

LG
TROMM

READER INPUT

Where makers tell their tales and offer praise, brickbats, and swell ideas.

» Skylar is 11 years old [pictured at right], and he loves to make things. He went to Home Depot and Lowe's three times this weekend. He is pretty tight with the plumbing dudes. They got a kick out of his weekend project. Not your typical DIY.

After many viewings of his fav podcast at makezine.com, Skylar made the T-shirt cannon this weekend. First launch was 7:40 p.m. He has already lost a T-shirt in the neighbor's bushes.

It works! Genius?

—Skylar's mom, Katie Wisdom Weinstein

» I happened across my first issue of MAKE magazine with Volume 05. At the time I was working on developing some children's alternative energy educational programs for a local startup in upstate New York. Your magazine sucked me in with the article about the DIY wind turbine from Velacreations [*page 90, "Wind Powered Generator," by Abe and Josie Connally*]. In a month's time I built my first Chispito wind turbine. From there it has been a whirlwind of change in my life.

I ended up building six Chispitos, became a forum administrator for the Connallys' website, traveled to southwest Texas to finally meet them in person in February 2007, decided two months later to sell my house in upstate New York and move to their part of Texas, designed an 8'×16' structure I could move into within two weeks, bought land in late November, began construction on Dec. 7, moved in on Dec. 19, and after ten weeks in the desert I'm just a couple weeks away from finishing all the details of my off-the-grid desert hut [pictured at lower right] — building materials $2,500, solar power system $1,500, water catchment system $500.

Funny thing is, I have yet to add a wind turbine to my place. This is just the first phase of a sustainable living field research facility. Will be posting the full set of construction photos at Flickr. Happy to email some to you — you gotta see it!

—John Wells
lifeoffthegrid@yahoo.com

The Beetlebot article in Volume 12 is a delightfully minimalist creation. I really appreciate how concise and symmetric the design is. I do have a suggestion, prompted mostly by a sense of missed symmetry.

If the polarity of one of the motors is reversed, along with a reversal of the NO and NC connections on that motor's switch, then both batteries will be used and drained "equally" — by that I mean that when the Beetlebot is running unobstructed, each battery will power one motor, and only when turning would all the power come from a single motor.

—Peter Langston

I love MAKE magazine; I have subscribed to it since the first issue. One thing I must say is that all other magazines that have online versions suck. They may have the articles themselves online, but nothing like yours. MAKE's is a pristine digital recreation of the print version, page by page. Good luck trying to find the article that was on page 89 in the 11th issue of another magazine. You might have luck searching for the title on their website, but when, and if, you find the article, the page is riddled with ads and popups. I don't remember getting *those* in my print edition. *—Jon Rutlen*

Praise for the "Upload" section (*MAKE, Volume 12*). Dude! This totally rocks! It's exactly the kind of thing that I'd been pining for (without being able actually to articulate the pine). If you ever spin this section off as its own publication, I will subscribe to this one, too.

—DeAnna Miller, subscriber to MAKE and CRAFT

I have been reading *Popular Mechanics* and *Popular Science* since the early 1930s. You are making me relive my youth, when you had to make your own everything. (Today, you buy everything.) I try to instill this in my grandchildren, but on to the computer they go. I am trying to teach them to marry a puppet to a computer, and we're still working on that. Well, you do a great job. And I, for one, appreciate it. Keep up the good work. *—Frank Chistolini*

First of all, I'd like to note that I am an avid fan of your magazine. The quality and clarity of instruction is, to my knowledge, completely unparalleled. This forces me to hold your publication to a higher standard than most, and after reading "Cutting Down a Tree" in Volume 12, I was compelled to write.

As an employee of the Parks Department of the State of New York, one of my chief duties is to oversee the takedown of diseased, dead, or otherwise dangerous trees. It is quite useful to have this skill, and I was glad to see it in MAKE; however, after a close reading of the article, I've found that the method described is much more dangerous than it could be, and is about as outdated as using a handsaw.

The problem is a matter of control of the tree as it falls, and the hinge described in the article is the instrument of that control.

I don't want to write an entirely new article in this already long letter, but bear with me for a minute. Think of a tree as it is coming down with the sharp, flat-bottomed notch described in the article. As the top swings down on the hinge, the top of the notch will almost always hit the bottom before the tree has swung more than 45° from the vertical.

If an appropriately sized hinge is in place with an incorrectly angled notch, it will always snap at this point, allowing the tree to go wherever it wants. A good grounding in physics is not necessary to understand that this is an untenable situation. The butt end — with the rest of the tree along for the ride — will kick straight back very quickly, roll off or kick off to one side, tip the tree in one direction or the other, or any other of numerous terrifying ways to be eating maple by the end of the process.

There is an easy fix for this, and that is to create a hinge that doesn't break either until the tree hits the ground, or until a great majority of its momentum is pointing straight down (which will not happen until the tree is well past the 45° mark).

The key is to make what is called an "open-faced notch" by cutting ⅓ of the way through the tree. This type of notch is a Pac-Man-shaped 90° angle split evenly along the horizontal — 45° above and 45° below. This will allow the tree to finish falling with complete control and land where you want it to land, i.e. not on your house, car, or self. Hopefully this information will prevent some tree-related injuries among your readers!

—Kevin Brew

MAKE AMENDS

Eagle-eyed reader Dave Bell pointed out a couple of "glaring" math errors in MAKE, Volume 07, "Home Molecular Genetics." On page 67, the directions should have read:

- **Air-dry the glob on the toothpick for 10 minutes, then scrape it into *0.75ml* (3 droplets) of running buffer and allow it to dissolve overnight at room temperature.**
- **Make a 0.02% solution of Methylene Blue in distilled water; some aquarium supply and pet stores carry Methylene Blue in 2.3%, so you'll need to dilute this at about *115:1*.**

AHA! Puzzle This

By Michael H. Pryor

MAKE's favorite puzzles. (When you're ready to check your answers, visit makezine.com/14/aha.)

River Crossing

A dysfunctional family has to cross the river. On one side of the river are the mom and 2 daughters, dad and 2 sons, the maid, and the dog. There is a boat only big enough to hold 2 people (counting the dog as 1 person). Only the adults (mom, dad, maid) are capable of operating the boat. Everyone has to get to the other side, without anything bad happening.

There are some difficulties. If the dog is left with anyone and the maid isn't there to control him, he'll bite. The dad can't stay with any of the daughters when the mom isn't there, or they'll get into an argument. Likewise, the mom can't be alone with either of the sons when the father isn't there.

Remember, only an adult can operate the boat (and it can't drive itself). How do they cross?

Cheesecake

An obsessive-compulsive ant starts at the top left-hand corner of a square, checkered tablecloth. He's trying to reach the bottom right corner with the cheesecake in it. He will only move at right angles following the grid lines of the tablecloth. Also, he will only move toward the cheesecake (in other words, down or to the right, never up or to the left).

If the tablecloth is a 2×2 square, the ant has 6 distinct paths he can take (see illustration). How many paths are there if he's on an 8×8 checkered tablecloth?

Michael Pryor is the co-founder and president of Fog Creek Software. He runs a technical interview site at techinterview.org.

Illustrations by Roy Doty

Go cuckoo for coconuts, munch on the ultimate brain food, turn day into night, and build your own greenhouse.

TOOLBOX

New LEDs Light the Way

LED Replacement Heads

$39–$66 surefire.com

Professionals — be they military, law enforcement, or fire/rescue — demand reliability and performance from their flashlights. Traditional high-performance torches used incandescent bulbs, and while never shying away from sheer lumen power, these bulbs were more than willing to consume lots of lithium batteries along the way.

Recognizing these facts, and also looking to expand their consumer base, SureFire recently unveiled a line of powerful LED replacement heads for their popular flashlights. The new P60L assembly and the KX1/KX2 replacement heads leverage Cree-based LED technology to provide maximum lumens with minimum power consumption.

The P60L reflector assembly converts any SureFire running a P60, P61, P90, or P91 incandescent lamp assembly into a longer-running (and higher-output in the case of the P60) LED flashlight. For a few seconds' work, SureFire states that the P60L drop-in will provide a maximum of 80 lumens with a runtime of up to 12 hours.

SureFire's KX1 conversion head replaces the entire incandescent head on their 1-cell E1E flashlight, and also replaces the older KL1 head found on earlier E1L models. Similarly, the KX2 head converts 2-cell E2E/E2D incandescents as well as older E2Ls.

With the introduction of these new assemblies, not only has SureFire brought the latest high-efficiency LED technologies to the masses, it also has introduced a practical, environmentally friendly lighting solution. *—Joseph Pasquini*

» Want more?

Check out our searchable online database of tips and tools at makezine.com/tnt.
Have a tool worth keeping in your toolbox? Let us know at toolbox@makezine.com.

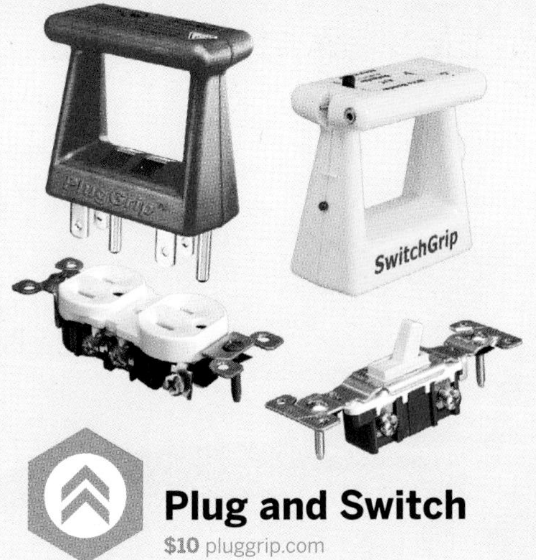

Plug and Switch

$10 pluggrip.com

The PlugGrip tool is used to install and remove electrical outlets. I had never heard of it until it caught my eye as I was picking out wall outlets for our remodel. This tool was amazing! It must have taken half the time to replace our outlets, and the tool even let us know that we'd wired them correctly, as its lights lit up when we turned the power back on. I would recommend this tool to anyone looking to replace or install new outlets.

The SwitchGrip, made by the same company, is used to install and remove light switches. We were just as amazed with it. The built-in wire cutter and bender worked really well. When we were done, we turned the power on and the SwitchGrip beeped to let us know that the switch was working.

—April Zamora

Lockdown

$12 loctiteproducts.com

Keeping fasteners fastened isn't always as easy as it seems, especially when it comes to vibration. Really cranking down on a nut or bolt won't prevent a fastener from coming loose, and you could end up damaging parts. Enter the world of thread-locking compounds: adhesives specifically designed to hold tight.

I recommend Loctite 248, aka "Loctite blue." Loctite has ingeniously utilized the glue stick as a vehicle for this thread-locker. Previously, Loctite was available only in liquid form, and while it worked well, its application was better suited to the bench than the field. The glue stick is great; it prevents waste and allows precise application of the material. Loctite 248 is a medium-strength thread-locker, meaning that you can undo what you've done without breaking your knuckles or resorting to a torch. I rarely fasten a nut or bolt without it.

—Alan Kalb

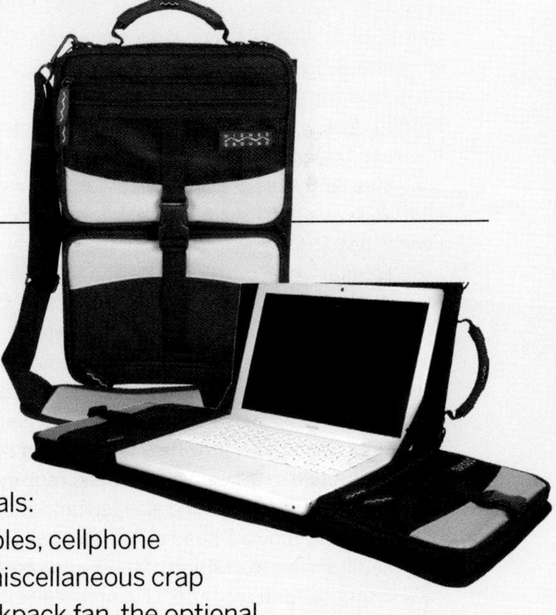

Laptrap

$60 highergroundgear.com

I've gone through over a dozen computer bags and had many favorites. But in recent years, I've stuck with the Laptrap. It's the smallest possible bag that fits all my essentials: 15" MacBook Pro, spare battery, charger, cables, cellphone power supplies, CDs and DVDs, and all the miscellaneous crap I put in my carry-on. And as a confirmed backpack fan, the optional $10 backpack strap sealed the deal.

One of the Laptrap's killer features is that the padded side pockets unsnap and fold open to turn it into an ad hoc lap desk. And I've come up with another way to use that feature: I can fit a change of clothes, folded flat, in the space beneath the side pockets.

—Brian Jepson

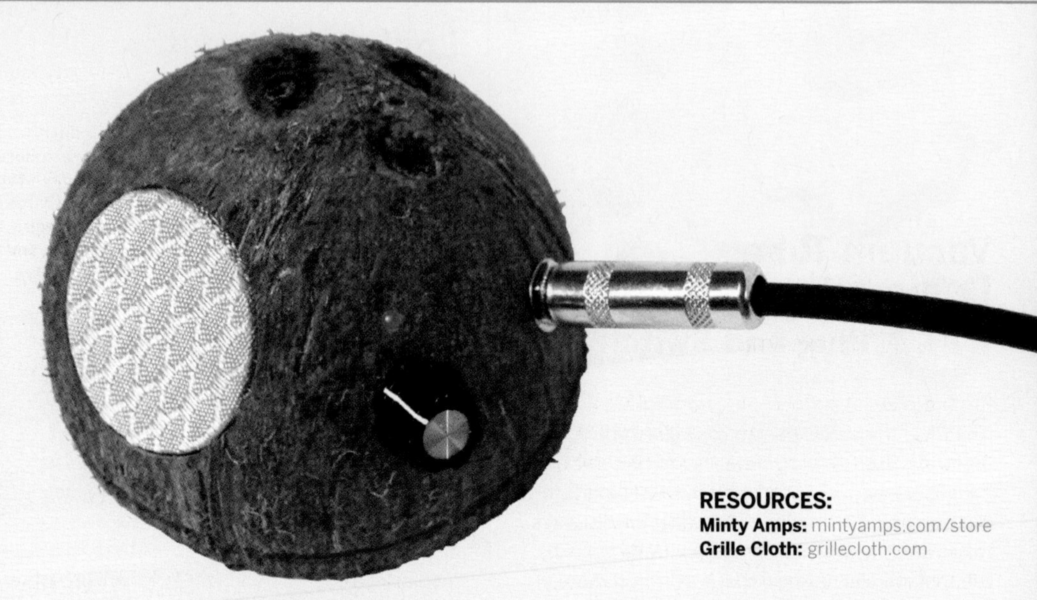

RESOURCES:
Minty Amps: mintyamps.com/store
Grille Cloth: grillecloth.com

Cuckoo for Coconuts

Although coconut shells do have some well-established niche market uses, such as novelty bikini tops and small caged-pet shelters, lately I've been feeling that too many of them are going to waste. So, while recently trying to find an enclosure for the ukulele amplifier I was building, I naturally reached for a coconut. I don't think I could have found a better case for my project.

In addition to the obvious thematic tie-in with the ukulele, the coconut shell has a number of other redeeming qualities. The shell is hard, durable, easily machined, and has a pleasing organic texture which can be left hairy, sanded smooth, or anything in between. The little brown dome of a coconut half-shell is cute as a bug, bringing a smile to all who see it. The parabolic shape is extremely stable and tip-resistant when used in the facedown position. Tip them up on edge and fashion a stand of some sort, and you'd have the basis for a really cool set of satellite speakers. I don't even want to think about how great a pair of coconut shell headphones might be.

For my coconut amp prototype, I opted to start with a basic amplifier PCB and parts kit from Minty Amps. (Building circuits from scratch using the parts in your box can be rewarding, but for speed and convenience, nothing beats a kit.) The Minty Amp circuit, built around the LM386 amplifier IC and a handful of other components, has a tiny footprint. Tiny enough to let you house your new practice amp, complete with battery, speaker, switch, inputs, and outputs, inside that favorite little project box, the mint tin.

The Altoids tin does make a nice project case. It has a hinged lid, giving you easy access to your business, a great pocketable form factor, and of course it's everywhere for free. The metal construction provides convenient RF shielding and a nice common electrical ground.

But despite these merits, I really had no choice but to opt for the coconut shell. I made a floor plate from ¼" plywood and secured it to the shell with 3 screws. The speaker grille is reproduction radio grille cloth stretched over a hoop made from a slice of PVC pipe. A classic-style knob for the power switch/volume control and a chrome input jack finish off the amp.

Now how are the Howells supposed to sleep with Gilligan making that confounded racket?!

—Steve Lodefink

COOL KITS FROM THE MAKER STORE

For these and more kits, visit store.makezine.com.

Vacuum Tube Radio Kit

$100 Product code: GKSSVTR

Put together a handsome vacuum tube radio from an easy-to-make kit. It includes a pin straightener for the vacuum tubes, a testing microphone so you can make sure everything is hooked up correctly to make sound, rubber feet on the fiberboard to minimize "howling," a variable condenser to allow for finer tuning, and a powerful transformer for volume and sound quality.

Photograph by Midel Toriyama / Mike Richards

Peggy LED Boards

$80 Product code: MKEVLPEG

Another evil project from Evil Mad Scientist Laboratories, this light-emitting pegboard display (affectionately known as "Peggy") provides a quick, easy, powerful, and efficient way to drive a lot of LEDs (up to 625) in a big matrix covering almost 1 square foot of area. You can make an LED sign for your window, a geeky valentine for your sweetie, one bad-ass birthday card, or freak the holy bejesus out of Boston. Your call. It's a versatile, high-brightness display. How you configure it and what you do with it is up to you.

Mechamo Kits

$60–80 Product codes: GKMKC-P, -R and -IW

Make metal inch, scuttle, and undulate with these satisfying kits. The final robots are as sculptural as art, but even more fun in motion. One remote works for all three kits, so you can have robotic insect wars between the inch worm and centipede, and then watch the crab crush them both!

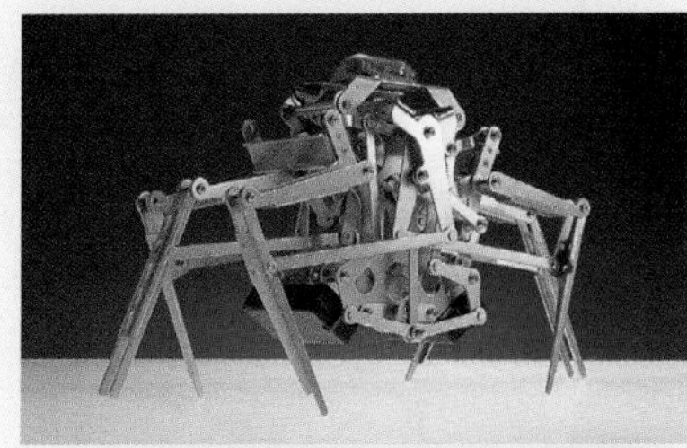

Telekinetic Pen Kit

$15 Product code: MKTELEPEN

From MAKE, Volume 13, comes this great DIY magic trick. Using a small battery, a magnetic reed switch, and a miniature pager motor, this device can be concealed inside a fine-point Sharpie pen. When the magnet is drawn near, the reed switch closes, making the Sharpie dance and vibrate as if by magic! Only you will know it's really just sufficiently advanced technology.

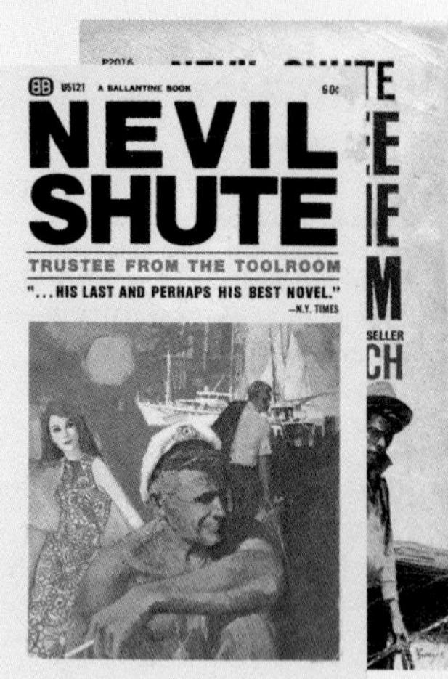

« Make Your Own Adventure

Trustee from the Toolroom **by Nevil Shute**

$16 House of Stratus

Sometimes good books of popular fiction just get lost. They're not extinct, just ignored into obscurity. This may be what happened to this gem. Set in the late 1950s, it's about an unassuming man who goes to extraordinary lengths to fulfill an obligation, taking him from a quiet suburb of London, to the South Pacific, to the Pacific Northwest, and back. Keith Stewart is a model maker who makes a modest income writing for *Miniature Mechanic*, a magazine "with a growing popularity amongst eccentric doctors, stockbrokers, and bank managers who just liked engineering but didn't know much about it." This simply written adventure is a terrific story where inventive model making stands Stewart in good stead on his odyssey.

—Kes Donahue

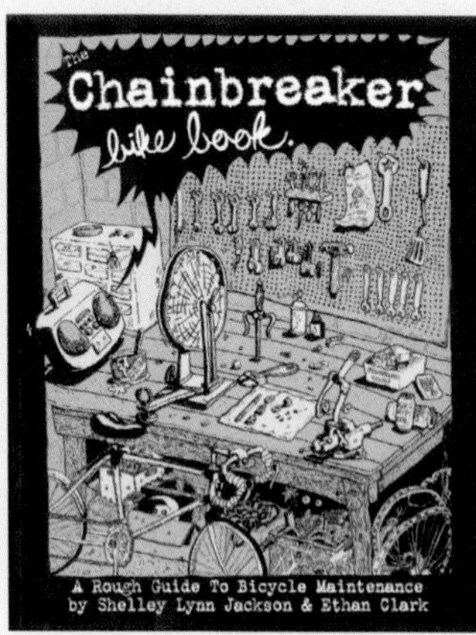

« Wheel Life Lessons

The Chainbreaker Bike Book **by Shelley Lynn Jackson and Ethan Clark**

$12 Microcosm Publishing

The first half of this book is about bike maintenance: how to choose a frame, how to build a bike, and how to fix it when it's broken. It overexplains some things, but it's the perfect intro for someone who's getting into bicycling and wants to learn more about their bike. (A lot of the women I know got into making stuff by building bikes, so consider this a gateway drug.) The back half of the book reprints four issues of *Chainbreaker*, a zine about bikes by Shelley Jackson. She started by publishing them out of her house in New Orleans, which was flooded after the levees broke. The originals were destroyed, but now you can see them in the book.

—Sam Murphy

« Gimme Shelter

Shelter **by Lloyd Kahn**

$25 Shelter Publications

This book is a structure addict's dream. From page 1, it's filled with stunning photographs, drawings, and charts from all over the world, showing how what we build to live in goes on to shape our lives. *Shelter* is an explosion of information: footnotes, articles, interviews, and other personal testimonies leak from the pages. How-tos are combined with even more outlandish but sustainable ideas from all over the world. Gypsy wagons, houseboats, tree dwellings, and location-specific habitations like the cliff debris villages of Timbuktu abound, on the scale from outsider culture to barn raisings to the building blocks of entire civilizations.

—Meara O'Reilly

As chosen by David Grosof of OptiOpia

« *Optical Designs in Motion*

***with Moiré Overlays* by Carol Belanger Grafton**

There's more to seeing than optics alone. This 1976 book of black-and-white optical art patterns comes with acetates that are also printed with op-art goodness. When the acetates are superimposed over the patterns and slowly slid, all kinds of interesting wavy-groovy moiré effects ensue, on page after page. Yet another reason recreational drugs are not necessary.

« Edmund Optics Catalog edmundoptics.com

We grew up with Edmund Scientific toys and kits, so we have a warm spot in our hearts for Edmund Optics. The catalog has amazingly lush and colorful photography of even the dullest mechanical widget, and good tutorial descriptions and diagrams that show how to use, say, penta prisms to turn an image 90° without inverting or reverting it.

True Vision glassesformissions.org

Reading glasses for 70 cents? No electricity required to fabricate? I wouldn't have believed it if I hadn't seen the glasses made in South Africa. These evangelical Christian missionaries think good vision equals access to scripture, but the tools work for everybody: check out this site to learn how to turn wire on a simple jig to make frames that hold inexpensive plastic lenses. The site could use a video and a wiki, but the ingenuity shines through. Clearly.

« Surplus Shed Catalog surplusshed.com

We look here for lenses, filters, mirrors, adapters, kits, subsystems, and components for making projects and concocting toys. We like the low-priced military surplus and odd-lot high-quality components for stimulating new ideas, and the kits and gadgets to tear to pieces. Our favorite find so far is a little combo: when you screw a focusing loupe magnifier into a pocket 8×20 focusing monocular you get a powerful microscope!

Moong Dal sukhis.com

Our local farmer's markets in the Bay Area sell the best crunchy-tasty snack ever: split, dehusked, and (I think) fried mung beans, called moong dal. At $5 a bag, it's a highly addictive and very yellow treat that enables one to skip lunch while making and playing.

« *Eyewear: Gli Occhiali* by Franca Acerenza

Whenever we get stuck, this book gets the creative juices flowing again. It surveys eyewear through the ages, with an amazing array of designs and exotic materials for monocles, pince-nez, sunglasses, and lorgnettes, from glasses for soldiers wearing gas masks to haute couture glasses for fancy folks. It's inspiring to see how many ways the problem of hanging lenses in front of the eyes has been solved.

OptiOpia, Inc., is a for-profit social business devoted to making eyeglasses and eyecare more affordable and accessible to the half-billion to billion people who need glasses and don't have them in developing countries. Check it out at optiopia.com.

Hoop Dreams

$39 hoopbenders.net

I've been a gardener for 26 years and never could afford a decent-sized greenhouse. When I first saw these benders on eBay, it seemed impossible that a tool priced so affordably could bend a steel hoop in 1 minute. Then I came across their website and bought one. The bottom line is that it really does the job and saves a lot of money. We bent the steel hoops for both our hoop greenhouses. This little tool will blow you away. I guess no one thought of a tool like this before because it's just too simple, and I love, love, love simple. *—Sholady, TNT contributor*

VoIP Magic

$39 magicjack.com

This PC device threatens to make expensive phone bills a thing of the past. I'm amazed at how well it works! It requires a broadband internet connection and can be used with any standard telephone — a real telephone number with area code is included. MagicJack is squaring off against the big boys in the VoIP market. Skype — probably the biggest competitor — also offers free calls, but only to other Skype users. The MagicJack's superior voice quality and easy setup are hard to beat. Installation is a snap: just plug the tiny device into any available USB port, then plug your telephone into the device and that's it! The computer (PC or Mac) automatically installs the software; no fussing with CDs or finicky drivers. When you lift the handset, you even hear a realistic dial tone. VoIP never sounded so good! *—Ken Delahoussaye*

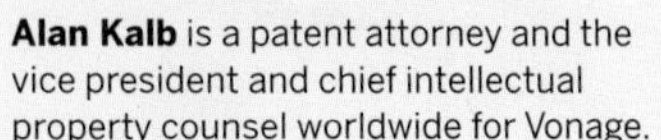

Alan Kalb is a patent attorney and the vice president and chief intellectual property counsel worldwide for Vonage.

April Zamora lives in Houston, Texas, where she works at a law firm.

Joseph Pasquini is an avid amateur radio operator and shortwave listener.

Ken Delahoussaye is a software consultant residing in Melbourne, Fla.

Kes Donahue is a rare books and special collections librarian at UCLA.

Meara O'Reilly is an intern for CRAFT magazine and a structure addict.

Steve Lodefink is a web designer and inveterate hobbyist in Seattle, Wash.

Tricks of the Trade By Tim Lillis

Take atmospheric night shots in broad daylight.

Need to shoot an interior shot during the day but want it to look like a night shot? Try this trick from George Cox at cogsindustries.com.

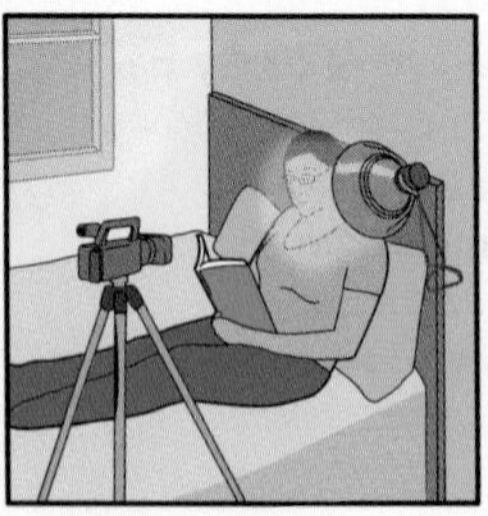

First, set up a light close to your subject. Don't worry about blocking out the windows.

Next, depending on your camera, either attach a neutral density filter to the lens, or enable the in-camera ND filter. You may want to use both.

Enjoy your quick and easy day-for-night shot and avoid messing around in post-production. For best results, experiment with your camera's ND filter settings.

Have a trick of the trade? Send it to tricks@makezine.com.

MAKER'S CALENDAR

Compiled by William Gurstelle

Our favorite events from around the world.

Jan	Feb	Mar	Apr
May	Jun	July	Aug
Sept	Oct	Nov	Dec

» JUNE

Texas Star Party

June 1–8, Fort Davis, Texas

At one of the top events on the amateur astronomer's calendar, star enthusiasts gaze heavenward under the dark skies of rural West Texas to observe and photograph the stars and planets. texasstarparty.org

Summer Explosives Camp

June 8–28, Rolla, Mo.

The Missouri University of Science and Technology provides high school students with a week-long opportunity to work with detonators, high explosives, blasting agents, display fireworks, and more. makezine.com/go/mst

Minnesota Inventors Congress

June 13–14, Redwood Falls, Minn.

"The world's oldest annual inventors convention" showcases ideas coming from the frenzied minds of noncorporate inventors. invent1.org

Fire Arts Festival

July 9–12, Oakland, Calif.

Makers with a burning desire celebrate at this annual event in the San Francisco Bay Area with classes, artworks, performances, and more. thecrucible.org/fireartsfestival

RoboGames

June 12–15, San Francisco

One of the world's largest robot competitions features events ranging from robotic combat to cybernetic bartending. robogames.net

» JULY

All-American Soap Box Derby

July 26, Akron, Ohio

Since 1934, this is the classic championship competition for young people and their handcrafted, gravity-powered coaster race cars. aasbd.com

EAA AirVenture

July 28–Aug. 3, Oshkosh, Wisc.

The biggest event on the aviation calendar takes place when more than 750,000 people attend the Experimental Aircraft Association's annual rally. airventure.org

» AUGUST

International Kite Festival

Aug. 18–24, Long Beach, Wash.

This is one of the largest and best-known gatherings of kite makers and flyers. More than 10,000 attendees revel in all things kite-related. worldkitemuseum.com

Bonneville Speed Week

Aug. 18–24, Bonneville Salt Flats, Utah

Gearheads and speed enthusiasts show their need for speed as they streak across the Great Salt Flats in a variety of fast-moving, high-powered vehicles. scta-bni.org

IMPORTANT: **All times, dates, locations, and events are subject to change. Verify all information before making plans to attend.**

Know an event that should be included? Send it to events@makezine.com. Sorry, it is not possible to list all submitted events in the magazine, but they will be listed online.

If you attend one of these events, please tell us about it at forums.makezine.com.

DRO FOR YOUR LATHES AND MILLS
$599 US
BT-2 System Includes
BT-2 Readout Unit
1 - 48" Scale
1 - 16" Scale
Power Supply
Manual
$749 US
BT-3 System Includes
BT-3 Readout Unit
1 - 48" Scale
2 - 16" Scale
Power Supply
Manual
Bolt Circles • Bolt Lines • Power Fail Memory • Tool Offsets
Reversible Count Direction • INC/ABS Mode • Inch/MM
Diameter Mode • 5 Year Warranty
Shooting Star TECHNOLOGY
CALL TOLL FREE
1-800-772-6322
www.star-techno.com
Made in Canada BUY FACTORY DIRECT

Precision CNC Machining
Your ideas shouldn't be limited by your tools! Whether it's new product development or just plain fun, a PCNC 1100 mill can transform your ideas into practical realities. Better keep it a secret though, everyone wants to be your friend once they learn the sort of capability you have.
3 Axis Mill
$6800
(plus shipping)
Shown with optional stand and accessories.
Tormach PCNC 1100 Features:
Table size 34" x 9.5"
1.5 hp spindle 4500 RPM computer controlled
Precision ground ballscrews
Optional 4th axis, digitizer, and more
Includes CAD and CAM software
Cuts aluminum, steel, even stainless!
TORMACH
Product information and online ordering at
www.tormach.com

Solderless clock kits
$25–33 with coupon 'mk5'
Free shipping!
Get started with our bright 9" x 5" kit
(LEDs, panel, 1 watt driver, stands, manual).
Then customize! We make it easy to:
Build even larger displays.
Synchronize multiple displays.
www.LEDkit.biz

Mold Making & Casting
Starter Kits
Smooth-On's Mold Rubbers and Liquid Plastics Enable You To Turn Your Ideas Into 3-D Reality. Ideal For Inventors, Designers, Artists, Engineers And For DIY Projects
Starter Kits contain everything you need to make a mold & casting.
SMOOTH-ON
www.smooth-on.com • 800.762.0744

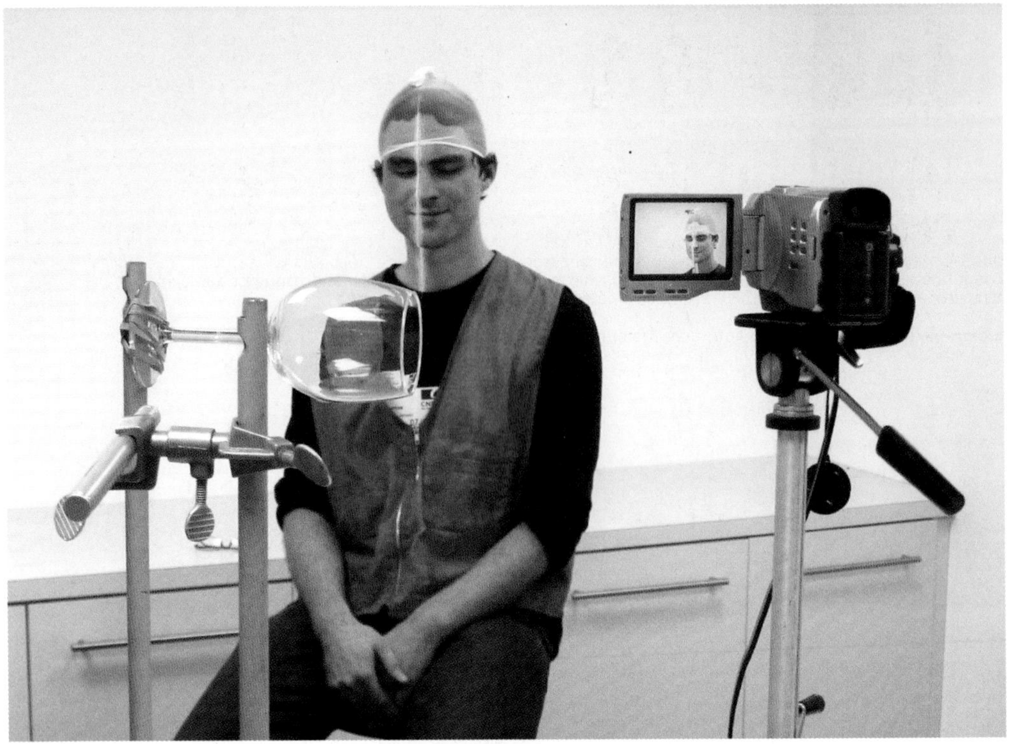

HOMEBREW

My Homebrew 3D Scanner

By Tim Anderson

Imagine a camera that captures the shapes of things, and a printer that prints out those shapes. We live in a world where paper is flat and computer displays are flat, but pretty much everything else is 3D. So how about taking photography into the third dimension? Not just as a visual illusion, but as a way to record three-dimensional shapes and then print them out as actual objects.

Years ago, friends and I got excited about this and built 3D scanners and 3D printers. We started a company called Z Corp. to make 3D printers, and these days there's a printer in every major town. 3D scanners are not so common, however, which is strange because you can build one in a few minutes.

This is Kenny Jensen, sitting on a stool that slowly rotates at a constant speed. It has an electric, motorized antenna rotor bolted to the seat post. A bubble level ensures that the axis of rotation is vertical. A laser pointer shines through the stem of a wineglass, which spreads the beam into a vertical line on his face. The wineglass stem is a cylindrical lens that spreads the beam in only one axis.

A camcorder looks at this green line from an angle and sees a wiggly, green line silhouette. The computer records the video frames of this wiggly green line. (The nylon sock holds his hair down. Hair is too fine for the beam, and shows up as spikes.)

How do we turn all those video frames of squiggly lines into a 3D model? Jensen wrote a MATLAB script that does it. (You can see the step-by-step instructions on Instructables at makezine.com/go/3Dscanner, including the script.) The software does some simple trigonometry to calculate the distance to each point on the surface of the face, then assembles the 3D model. There's commercial software that does the same thing, as well as a rapidly growing collection of free utilities.

Once I had 3D models of myself, I proceeded to 3D-print them and then cast those parts in metal. I made some titanium toy soldiers of myself. I also made self-portrait cremation urn bookends of my head. I thought they were pretty creepy, but my mom loved them. She's still got them on her mantel.

Video of the scanner in action: makezine.com/go/3Dscannervid

Tim Anderson is a columnist for MAKE and a founder of Z Corp.

Photograph by Tim Anderson